574.072 Lewis, Alvin
LEW   Edward, 1916-

Biostatistics

1966

2495

DATE			

BIOSTATISTICS

SECOND EDITION

BIOSTATISTICS

SECOND EDITION

Alvin E. Lewis

1966

VNR VAN NOSTRAND REINHOLD COMPANY
NEW YORK   CINCINNATI   TORONTO   LONDON   MELBOURNE

Library of Congress Catalog Card Number: 83-16744
ISBN: 0-442-25954-9

Manufactured in the United States of America

Published by Van Nostrand Reinhold Company Inc.
135 West 50th Street
New York, New York 10020

Van Nostrand Reinhold Company Limited
Molly Millars Lane
Wokingham, Berkshire RG11 2PY, England

Van Nostrand Reinhold
480 Latrobe Street
Melbourne, Victoria 3000, Australia

Macmillan of Canada
Division of Gage Publishing Limited
164 Commander Boulevard
Agincourt, Ontario M1S 3C7, Canada

15 14 13 12 11 10 9 8 7 6 5 4 3 2 1

Library of Congress Cataloging in Publication Data

Lewis, Alvin Edward, 1916–
    Biostatistics.

    Bibliography: p.
    Includes index.
    1. Biometry.   I. Title.
QH323.5.L47 1984   574′.072   83-16744
ISBN 0-442-25954-9

CONSULTING EDITOR'S STATEMENT TO THE FIRST EDITION

One of the difficulties in teaching contemporary biology is to find a statistics text that joins biology to mathematics in a marriage of love, rather than of convenience. Dr. Lewis has been outstandingly successful in this difficult task for two reasons. First, he has based the book on sound mathematical principles and, second, he has copiously illustrated these principles with the type of biological data that belongs to today. The book gets right down to fundamentals by discussing the relation between probability and randomization, and then deals with random selection and the description of data. A satisfactorily brief chapter on the "normal" curve leads first to sampling and the universe of discourse, and then to a consideration of the null hypothesis. The student is then taken smoothly through regression, correlation, enumeration statistics, Chi square, and the analysis of variance to a thorough discussion of tolerance limits, including a chapter, rare in biology statistics books, on quality control and its application to biological problems. The book comes to an end with discussions of alternatives to the null hypothesis, sequential analysis, and a chapter on nonparametric statistics. Each of the sixteen chapters has at its end, for ready reference, a summary of the chapter. The appendix has an unusually full set of those tables necessary to the working statistician.

This book is what I have long wanted for my own course in biostatistics so that I welcome it as an admirable newcomer to the REINHOLD BOOKS IN THE BIOLOGICAL SCIENCES, not only as an editor, but also as a teacher.

PETER GRAY

PREFACE TO THE SECOND EDITION

When the first edition of this book was written over 15 years ago, several topics were included that seemed useful and important for biological investigations. As the years passed there was little need to use them either directly or in consultation, and our view of their importance has been drastically revised. At the same time topics we then regarded as less useful we have now come to consider as essential.

In spite of these misguided enthusiasms the original edition and its Spanish translation continue to be used. It is gratifying then to have an opportunity to make important improvements in this second edition. This introductory text should now be more directly helpful in the evaluation of medical and biological data and in the planning of investigations.

For all that has been done by colleagues, friends, and loved ones to help me, my gratitude is undiminished by the passage of time. My effort here to provide a lucid introduction to biological statistics is in part an expression of my appreciation.

A. E. LEWIS

PREFACE TO FIRST EDITION

If this were the first book ever written on biological statistics, this author's task would have been a simple one. There are, to be sure, a remarkable number of texts on this subject as well as on the subject of statistics in general. These vary from simple formularies with instructions covering typical applications to sophisticated treatises on statistical mathematics. To a degree these would appear to match the particular needs of specific groups of students with varying levels of mathematical skill. However, in spite of the wealth of available choices, a substantial gap remains which, hopefully, this book will fill.

This book is designed for students in biology and medicine who have reached the stage where they are ready to judge data and to begin their own investigations and experiments. It is assumed that these readers are able to follow ordinary algebraic manipulations. A background that includes calculus would guarantee an adequate mastery of the necessary algebra, but calculus is not required for this text. On the other hand, the student who has had no algebra beyond the high school level and has avoided using even this will find some portions of this book a little tedious.

Relatively little mathematical skill is demanded of the reader, but at the same time the aim here is to provide sufficient insight into the processes of statistical analysis to use them intelligently. The book is intended to take the student beyond the usual introductory exposition of the t-test, χ^2, regression, and correlation. Particular care has been given to the exposition of the analysis of variance; even if this is not included in a one semester course, the student may need it later in his career for the design of experiments. Other items presented that are not usually included in these elementary courses are quality control as applied to biological and clinical investigations, some nonparametric methods, elements of sequential analysis, and hypotheses testing with particular emphasis on their relationship to determinations of sample size. This text should continue to serve the student long after the completion of formal course work.

Wherever possible the algebraic developments have been linked to situations. Since these expositions rely heavily on intuitive reasoning, they are often lacking in either mathematical elegance or rigor. However, this book is not designed for mathematicians, but for biologists and clinicians who otherwise would not find this material in a form easily intelligible to them.

For whatever merit this book may possess I am indebted beyond measure to those who have taught me. My interest in this subject was initiated by the late F. W. Weymouth who along with John Field, II and Victor E. Hall guided my first struggles with biological measurements in the Physiology Department at Stanford University. For the courage to look further into the theoretical basis of measurement, I must thank my former colleagues at U.C.L.A., Moses A. Greenfield and Amos Norman. Their friendly and informal discussions ranging from games theory to mutation rates made me aware of the role of probability theory in areas outside the confines of my own special interests.

Most of this book was written during the time that I was responsible for the operations of the clinical laboratories and for the training of residents in Clinical Pathology at the Mount Zion Hospital and Medical Center in San Francisco. I shall always gratefully remember the challenges and kindly encouragement of my colleagues and students at this institution.

I am indebted to the Literary Executor of the late Sir Ronald A. Fisher, F.R.S., Cambridge, to Dr. Frank Yates, F.R.S., Rothamsted, and to Messrs. Oliver & Boyd Ltd., for permission to reprint Tables Nos. A2, A3, A4, A7, and A8 from their book *Statistical Tables for Biological, Agricultural and Medical Research*. I am indebted to the RAND Corporation of Santa Monica, California for permission to reprint the material in Table No. A1 from their publication, *A Million Random Digits*. I am also indebted to D. L. Burkholder, Editor of *The Annals of Mathematical Statistics* for permission to reprint Table A5 from "Tabulated Values for Rank Correlation," by E. G. Olds, appearing in Volume IX of the Annals for 1938.

Just as actors need a producer and a stage to reach their audience, an author needs a publisher to provide the management of the complex operations of the publishing industry. The staff of the Reinhold Publishing Corporation has been most helpful with this venture from its inception. I am particularly grateful to Mr. Leonard Roberts, Editor, for his stimulating encouragement and to Mrs. Cynthia Harris, Copy Editorial Supervisor, for her patience and for her meticulous review of the manuscript.

Finally, I must pay tribute to my wife and to my daughters. Without their patience, understanding, and encouragement this project would have ended with Chapter I. This book is dedicated, then, to Doris, Joan, and Elizabeth from an affectionate husband and father as a small measure of compensation of the time together that we have lost forever.

April, 1966 ALVIN E. LEWIS

CONTENTS

Preface to the Second Edition   vii

Preface to the First Edition   ix

1. Introduction   1

2. The Relationship of Probability to Randomization   6

3. Random Sampling and the Description of Data   19

4. The Normal Distribution Curve   29

5. Samples and the Universe of Discourse   45

6. The Null Hypothesis and Comparison of Means   55

7. Graphs and Equations (Regression)   66

8. Correlation   84

9. Enumeration Statistics   95

10. The Poisson Distribution   112

11. The Chi Square Distribution and Variance Ratios   116

12. Comparing Proportions in Small Samples   125

13. Analysis of Variance   131

14. Quality Control   145

15. Testing Alternatives to the Null Hypothesis   154

16. Distribution-Free Methods (Nonparametric Statistics)   162

Epilogue   170

Answers   173

Appendix   177

Index   195

BIOSTATISTICS
SECOND EDITION

1
INTRODUCTION

While the mathematical theory of statistical methods is not usually part of the standard equipment of students in biology, statistical reasoning and techniques can be mastered and applied by anyone with a fair grasp of algebra. Nothing would be gained by suggesting that these methods are easy. They do require thought and application, but their inherent interest and obvious utility can make them reasonably pleasant and satisfying.

The subleties of statistical reasoning often escape the student in his first encounter with statistical methods. Obtaining the answer to textbook problems is only the first step in mastering statistics. Mystique has no place in science, but in an elementary exposition of this kind intuitive understanding will have to be substituted for several years of mathematical preparation. The purpose of this text is to aid the student in acquiring a useful mastery of the subject by drawing wherever possible on the commonsense experience of daily judgements.

Consider first the well-known average and some of its ordinary uses. For example, suppose we need to order a ten day supply of food for 100 experimental animals. If the average requirement per animal for ten days is known, all we have to do is multiply this amount by 100. Even though some animals eat much more than the average, we do not need much statistical knowledge to feel reasonably sure that the large eaters will be balanced by an almost equal number of small eaters. On the other hand, if instead of 100 animals, we have to feed only one or two, we would not be surprised if the average amount of food turned out to be either excessive or inadequate. This obvious example gives an intuitive basis for making a few useful general statements.

First, there is the average itself. If the average food supply should be given in a handbook or manual, we would assume that the value reported is a summary of some comparable experience with a large group of animals, although this assumption might not be made consciously. This common average is more properly called an *arithmetic mean* in the language of statistics. As shall be pointed out later, there is also a *geometric* and a *harmonic mean*. Each of these is quite different from the arithmetic mean. Generally, when the term *mean* is used without a modifier, the arithmetic mean or average is implied.

The mean, then, regardless of type, summarizes in a single number many individual values. A summarizing value of this kind is necessary, because the human mind is unable to grasp multiple impressions simultaneously. Again, we intuitively or unconsciously assume that a mean or average is significant and useful only when it summarizes a large number of values. We assume that the average food requirement in the handbook was obtained on the basis of a large number of animals and do not expect to apply this value precisely unless we too are dealing with a large number of comparable animals.

So far, most of the discussion has been an elaboration of the obvious; this is necessary perhaps, but fairly obvious all the same. One assumption was noted, however, which may not be obvious and in spite of our intentions may not always be true. This assumption is that each animal eating a certain amount more than the average would be matched by another in the group eating almost the same amount less than the average. In other words, the animals could probably be paired off so that the average for each pair would equal the average for the group. This implies that the frequencies of each value are symmetrically distributed on either side of the mean.

Another notion, which is readily accepted, concerns the range of food requirements. The range of requirements is the difference between the largest and the smallest values in the group. Without any intensive thought, or recall of past experience, most of us would readily agree that extremes in food requirement are unusual. That is, most of the animals would have requirements not too different from the mean. The extremes at either end of the scale represent a small minority.

So far, the discussion has been an intuitive, qualitative description of a common or *normal frequency distribution*. Symmetrical frequency distributions with the majority clustered about the mean occur in numerous instances. In the study of statistical analysis these concepts will be used quantitatively.

Up to this point, animal dietary requirements have been used as an example of the average and the distribution of values of a *continuous variable*. There are many kinds of measurements that fall into this category. These are measurements that increase by vanishingly small amounts, the smallness being limited by one's ability to discriminate correspondingly fine differences. For example, if we were to take 1000 people, all of them weighing between 150 and 151 pounds, we could, if the scales were sufficiently sensitive, arrange them in order of increasing weight. Obviously, in order to do this, we would need a scale that could discriminate differences smaller than 1/1000 of a pound. Theoretically, with an unlimited population to draw from, we could take 1000 people weighing between 150.002 and 150.003 pounds and also arrange them in order of weight if the scales were sensitive to less than 1/1,000,000 of a pound. Needless to say, in real life there would never be any occasion to carry these measurements to such hair splitting accuracy. Nevertheless, a mean value of, say, 150.01 pounds has conceptual reality.

On the other hand, if a statistician states that the average family has 2.3 children, we balk at the image of three tenths of a child. We do not for a moment deny the utility of this mean for certain economic purposes, but we can immediately perceive that another class of values is involved. These are called *discontinuous variables*. They are obtained by counting or simple enumeration rather than by measuring against a scale of some kind. In genetics we *count* progeny with distinctive characteristics; in studying epidemics, we *count* cases; in bacteriology we *count* organisms. In all of these examples the units are indivisible. The count moves up discontinuous steps instead of rolling up a continuous slope.

Our commonsense, intuitive grasp of chance and counting again comes to our aid in understanding the statistical behavior of discontinuous variates. Indeed, our grasp of probability theory, which is needed in order to make statistical analysis useful, has its roots in our intuitive notions about such variates. A simple example will suffice for now, and although superficially it may appear trivial, the same example will be explored in considerable depth in a later chapter.

Suppose we take 100 coins, shake them in a box, and see how many turn up heads and how many tails. Without actually doing this experiment, we expect close to 50 heads and 50 tails. Remember that in performing this task, we would find the number of heads by counting, not by measuring. There might be 47 heads instead of the expected 50 but obviously there will not be 47.4 heads. The important intuitive point here is that we *expect* close to 50 heads. We would be quite astonished if all of the coins turned up heads. We would, a priori, (i.e., before any actual experience) state with confidence that such an outcome was *improbable*.

Notice that, being careful scientists, we did not say impossible. As a matter of fact, we realize that if the experiment were repeated over and over again an enormous number of times, eventually 100 heads would turn up simultaneously. We do accept the inevitability of an improbable event given enough trials. While 100 heads turning up at the same time might be regarded as a near miracle, biologists in general accept the more nearly miraculous events of evolution as purely fortuitous.

In these introductory comments we have introduced some new terms, or more likely, we have introduced some old words in a new light. We have not attempted any rigid definitions so far. For pedagogic reasons, whenever it is possible, precise definitions will be given as summaries after the sense of a discussion has made them useful.

Before these introductory comments are concluded, the meaning of the words *statistics* and *probability* will be considered. Statistics may be defined as the science and technique of gathering, analyzing, summarizing data, and estimating the probability of inferences from these data. We shall occasionally use the singular form, statistic, to refer to a value either observed or generated in a random manner.

Probability is a word that might better be used here without definition, as its meaning is still a matter of active philosophic debate. The word shall be used as if it meant the long run relative frequency of multiple or repeated events.

With statistics and probability defined adequately for present purposes, it might be useful at this point to indicate some of the things that statistical analysis is not and to point out the areas where our technical concept of probability does not apply. Experimenters should be reminded occasionally that statistical analysis is not a magic way of converting poor data into good data. Statistical analysis helps one deal with random variability produced by unknown, small causes, but it helps little, if at all, in dealing with variability due to poor or inadequate experimental technique.

In ordinary conversation we frequently use the words "probable" or "probably." For example, we might say that Peru and Guatemala will probably sign a treaty tomorrow. Notice that there is no way of giving a meaningful fractional or percentage value to this probability, since the situation is not representative of a multiple or repeated event. Another example of the way in which probability may be misinterpreted is as follows: suppose 100 students apply for admission to a graduate department which has only 25 openings. At first, we might think that a friend has a one in four chance of being accepted and might even make a bet giving odds one way or the other in the same one to four proportion. This would be ignoring the obvious fact that although the friend would either be admitted or not be admitted, the outcome would not be determined merely by random chance. Random chance would operate only on the probability of our winning bets if, in total ignorance of the factors determining entry, we made the same bet with many members of the group applying for admission. Insurance companies and bookmakers operate on this same principle of safety in numbers, but in a nonrepetitive instance or in the individual case, probability statements may have little or no valid meaning.

SUMMARY

Statistics may be defined as the science and technique of gathering, analyzing, and making inferences from data. These inferences are stated as probabilities. In this connection the term "probability" is used as if it meant the long run relative frequency of multiple or repeated events. The data subjected to statistical analysis consists of two types of variables: (1) continuous and (2) discontinuous. The former consist of measurements against a scale while the latter consist of data obtained by counting or enumerating discrete, indivisible units.

2
THE RELATIONSHIP OF PROBABILITY
TO RANDOMIZATION

We usually have a clear definition in mind, which is readily understood by others, when using the words or expressions "pure chance,"
"probability," and "random." However, some difficulty is encountered when these ideas are examined as closely as is necessary for
their utilization in making calculations. The most common example
of pure chance is the toss of a coin. Whenever we prefer to let
fate make a decision, we toss a coin. The first official act at the start
of a football or baseball game is the tossing of a coin.

Generally, we accept the notion that heads will occur about as
often as tails and thus that each side has an equal chance of being
favored. At least we feel that "in the long run" the equality will
hold. The important point to notice is the implied condition of a
long run. A sequence or run of three or four heads in succession
would not be surprising. During a trial of ten coin tosses performed
while this paragraph was written, the following sequence occurred.

HTTTTHTTTHT

However, our confidence is still unshaken, and we continue to expect
the proportion of heads to tails to even out if the run is sufficiently
prolonged. The basis for this confidence lies in the symmetry of the
coin.

In contrast, consider expectations with a coin flipping machine
constructed as shown in Figure 2-1. If the coin is always placed
snugly against the back stop of the pedestal, if the flipper always
delivers the same amount of force, and if the coin is always placed in
the heads up position, we would expect the coin to show the same
face each time it is tossed. Thus, if the coil came up tails on the first

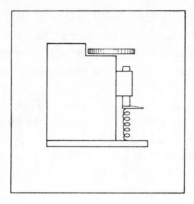

Figure 2-1. Mechanical coin tosser.

that, given the dimensions of the apparatus, the weight and diameter of the coin, and the force of the flipper, an engineer skilled in analytic mechanics could predict the outcome without any experimental trial at all.

Suppose, by way of contrast, we attempt to duplicate the machine's consistency by hand. We would immediately become aware of the many small but almost uncontrollable adjustments that would be necessary to make our flipping techniques competitive with the machine. We have already agreed that when the controlling factors are known, the outcome of an individual toss is known. Therefore, we must agree that the difference between a completely predictable outcome and a random outcome is our inability to control or evaluate the numerous, small factors that determine that outcome. In essence then, at least as far as coin tossing is concerned, the operative and determining factors in one toss may or may not be dominant in the next toss. With the machine the results show complete bias; when we try to imitate the machine by hand, incomplete but fairly substantial bias is introduced; when we simply toss the coin without any attempt to control it (i.e., at random), any remaining bias must be negligibly small.

Thus, *random* may be defined in a negative way as absence of bias or of factors known to contribute significantly to the outcome of any trial. To stress the more positive approach, we may consider a result to be random when the outcome is determined by the inconsistent interplay of many small factors.

A slightly more complicated but highly useful example is found in the classic white ball-black ball problems. Suppose we have in an urn 1000 balls, which are physically identical except that 100 of them are black and 900 are white. The container is shaken so that the balls move about at random. Theoretically, it would be possible to calculate the locations of each of these balls if all the starting positions, weights, force vectors, etc., were known. Obviously, this calculation would be too enormously complex to make it worthwhile; but the final location of each ball would be the result of the interplay of many small, virtually incalculable forces. In short, the balls would be distributed *randomly*.

Suppose further that ten balls are withdrawn by a blindfolded experimenter and the number of black balls in the sample recorded. The balls are returned to the container, shaken up (i.e., the balls are randomly redistributed) and the experiment repeated. This process is carried out over and over again. Just as in the case of the coin tossing experiment, we are able to predict *a priori* that in the long run, one out of every ten balls sampled will, on the average, be black. Similarly, remembering the definition at the end of Chapter I, we would say that the probability of selecting a black ball by chance from the container is one in ten, or more concisely, 1/10, as we would say of the tossed coin that the probability of heads turning up is 1/2.

To emphasize the lesson a little more strongly, consider, for example, a bag containing only five balls; one is white, and the other four are black. The experimenter shakes the bag without looking and draws a ball at random. Its color is noted, and the ball is returned to the bag. Repeating this procedure over and over again, we would say, *a priori,* that the probability of selecting a white ball by chance is 1/5. Thus, the principal element in predictions of this kind is not the large or small number of balls used; it is the large number of trials, or in other words, it is the *long run*.

The essential point of the foregoing examples is that we can state *a priori* the value of the pertinent probability when we know all the possibilities that can occur at random. In real life, particularly in the biological sciences, we rarely have such complete knowledge. Instead of predicting probabilities ahead of time from a knowledge of all the facts, we have the more difficult task of inferring some fragments of knowledge from observed proportions in the available data.

Returning now to the urn with a large number of balls, suppose we knew only that there were some black balls among the white balls but had no idea of the proportion. From long run data accumulated as above, we could infer quite accurately that 1/10 of the balls were black. Notice how *inference* is just the reverse of *prediction.* Formerly, we knew the proportion of black balls and *predicted* the probability or proportion in the samples. Here, we observed the proportion experimentally and from this information *inferred* the proportion in the urn. Statistically, in the latter instance we used the *sample population* to measure the *parent population.*

In one way or another statistics are used for making reasonably accurate measurements of a parent population. The subject matter of statistical analysis is concerned not only with the problem of making these measurements but also with the problem of estimating the validity and accuracy of these values. The urn containing balls of one color or another or the coins that we toss serve as statistical models for problems far less trivial. Despite the convenience and simplicity of these models, we must not overlook the critically important features implied in the use of such models. The models must be relevant; the sampling must be random and free of bias, and the number of observations must be sufficiently large.

Suppose, for example, that we wish to determine the proportion of the insects in an area that are potential carriers of a certain disease. We could determine this exactly by trapping and examining *all* of the insects. Since this is obviously impractical, we must rely on an observation of the proportion of carriers in a suitable sample. If the carrier state has no effect on insect behavior, we may expect the parent population of insects to randomize their location in space as effectively as balls shaken in an urn. A hundred or more insects trapped and examined would ordinarily constitute a representative sample. As will be seen in a later chapter, the adequacy of the sample size is determined by the value we seek to measure and the accuracy that we demand of the statistical inference.

Frequently, a parent population is not obliging about providing ready made randomization. The observer must then provide randomization by one of many possible techniques. One illustration will suffice for the present. Suppose we wish to determine what proportion of the population in a large city has defective vision. Examining each and every inhabitant, would give us an exact answer, but the value or need for the information would not be proportional to the expense and effort of a 100% survey. We can obtain a satisfactory estimate of the proportion by taking a representative sample, but we must be certain that the sample is a random one. We may be unable to place the citizens into a gigantic urn, but if each citizen is represented by a numbered bean in an urn, a sample of randomized beans can be taken to select a representative random sample of citizens for examination. This is precisely the technique that has been used on various occasions to select draftees for the armed forces. Random chance is used in these instances to eliminate bias on the basis of social or financial status.

Although randomization is necessary before any sample may be considered representative of a parent population, this does not guarantee that the proportion in the sample matches the true proportion existing in the parent population. As pointed out previously, we intuitively realize that the larger a sample is, the more likely it is to be representative. This relatively simple notion about sample size does not put us in any better position than the scientists who worked before the development of statistical methods. These men who were careful and conscientious about their researches repeated their experiments and observations over and over and over again. However, such repetition is not always possible, and even large samples or large numbers of observations do not by their size alone dispel all uncertainty. There are, as will be seen later, some situations where even fairly enormous samples yield inadequate data.

Analytic considerations of some aspects of a priori probability supply us with a rational basis for solving this problem. Return now to the simple situation of a randomly tossed coin. Since it is symmetrically constructed and relatively flat, half the time it will, in the long run, turn up heads, half the time tails. Suppose that instead of tossing one coin, we toss two simultaneously. If the coins are numbered as coin No. 1 and coin No. 2, we can see that on half of the occasions when coin 1 comes up heads, coin 2 will also turn up heads. But coin 1 turns up heads in only half of the tosses. Therefore, the simultaneous appearance of heads with coin 2 occurs half of half or one fourth of the time. Table 2-1 summarizes the possible combinations.

If we wish to obtain the number of possible results using three coins, we can determine this by enlarging the table. For each of the four possible results indicated for two coins the third coin can fall

Table 2-1

No. 1	No. 2	Result
H	H	H H
	T	H T
T	H	T H
	T	T T

with either heads or tails up. In other words, when another coin is added to the group, the number of possible results is doubled. Similarly, if we add a fourth coin, the number of possible results is doubled again. Thus, with three coins there are eight possible results, while with four coins there are sixteen possible results.

Table 2-2 shows one way of illustrating the proliferation of possibilities as the number of coins is increased to three. A somewhat more

Table 2-2

No. 1	No. 2	No. 3	Combination
H	H	H	H H H
		T	H H T
	T	H	H T H
		T	H T T
T	H	H	T H H
		T	T H T
	T	H	T T H
		T	T T T

useful way of picturing this multiplication of possibilities is illustrated by the "tree diagram" shown just below the chart. Whether tables or diagrams are used is not important. These are merely devices to illustrate, in a concrete way, the abstract principle known as the *multiplication principle.*

The *multiplication principle* states that if a result or procedure can occur in a given number of ways, n_1, and if another result or procedure can occur independently in a number of ways, n_2, the number of possible combinations of results or procedures is the product of these two numbers, $n_1 n_2$. For each additional result or procedure to be combined with the preceding combinations, this product is multiplied again by the number of additional choices. Thus, for example, if we wish to calculate how many four digit license plates can be made without duplication, we see that the first space can be filled in ten different ways using the numbers 1 to 9 and including 0 as a number. Each of the subsequent spaces can be filled in an equal number of ways. Then the total number of plates that can be made is $10 \times 10 \times 10 \times 10$ or 10^4 (i.e., 10,000). Note that this does not include the total number of plates that can also be made with less than four digits.

The application of the multiplication principle can be illustrated in a little more detail if we consider the apparently irrational insistence of the telephone company that better service can be offered by switching to all digit dialing. Before the change the telephone numbers in large communities were all made up of two call letters followed by five digits. Since there are 26 letters in the alphabet, it might seem at first that the total number of separate lines that could be individually identified and serviced with the present equipment would be $26^2 \times 10^5$, or 41,600,000. The fact is, however, that there are only ten locations on the dial of the telephone, and since the letters Q and Z are not used, the 24 remaining letters appear on the dial in only eight groups of three, adjacent to the numbers 2 through 9. Thus, with the system of two call letters and five digits the maximum number of separate phone numbers would be 6,400,000. With an all digit system the maximum number is 10^7 or 10,000,000. This represents a gain of 3,600,000 in the seven digit system alone. Then if there are 1000 times as many when a three digit area code number is added, we can see how large an increase is possible without excessive modifications or additions to the equipment now in use.

Returning again to the simpler problem of tossed coins, we can generalize our understanding of the situation up to this point by stating that if n coins are tossed, there will be 2^n possible results. The reader will realize, however, that since the coins are not numbered or easily distinguished one from the other, many of the possible results will be indistinguishable. Thus, for example, with four coins the outcome, H T H T, would be regarded simply as two heads and two tails. This would not be distinguished from the possible result, H H T T.

In Table 2-2 the number of such otherwise indistinguishable combinations is summarized for three coins. The total number of possible combinations is eight, and we intuitively grasp the notion that each of these possibilities is equally probable. That is, in the long run each of these possible combinations would occur an equal number of times. Now if the results are summarized as shown in Table 2-3, we can see from the foregoing considerations that, in the long run, the combination of two heads and one tail may be expected 3/8 of the time. Since probability has been defined as an expression of the proportion of the time that a certain result can be expected, we say that the *probability* of obtaining two heads and one tail when three coins are tossed at random is 3/8. Although the probability here is expressed as a common fraction, probability is just as often expressed as a decimal fraction, 0.375, or as a percent, 37.5%. The choice is merely a matter of convenience, and these forms may be used interchangeably.

As has already been suggested, we can, if we wish, make diagrams or tables larger to cover the possible combinations with four or more coins, and then we can count the number of times a given distinct combination (e.g., three heads and one tail) occurs out of the total of

Table 2-3

No. of Heads	No. of Tails	Times
3	0	1
2	1	3
1	2	3
0	3	1

2^4, or 16, possibilities. Obviously, this procedure soon becomes impossibly cumbersome as the number of coins increased. Thus, if we wanted to know the probability of obtaining three heads and seven tails when ten coins are tossed, we would have to make a diagram showing 2^{10} or 1024 possibilities. Fortunately, mathematicians have provided us with some useful rules for accomplishing this task. In a later chapter we shall learn to solve problems of this kind using the binomial expansion. For present purposes, however, we shall make use of a more primitive system using *Pascal's triangle*. This numerical pattern is shown in Table 2-4. One can readily see by inspection the manner in which this pattern of numbers is generated. Except for the number one's at the outer edge, each number is the sum of the two numbers just above, as indicated by the inserted triangle. Even the number one's along either margin may be thought of as arising from the addition of the one above and a zero which is implied but not shown.

Of course, the inherent biological interest in coins in negligible, to say the least, but the paper experiment summarized in the table takes on new interest if we think of the coins as a statistical model. In a previous discussion, the sampling of a population for the selection of draftees was randomized with numbered beans. In a similar way we can use the behavior of coins to represent a phenomenon of some biological interest.

For this purpose a completely hypothetical situation will be used, but the possible connection with real situations should be fairly obvious. Assume that there is a species of animal whose size in grams above a certain minimum is determined by the presence or absence of ten independent but equal factors. These factors may be labeled alphabetically from A through J. If all of the factors are present as represented by the pattern ABCDEFGHIJ, the animal will be 10 g

Table 2-4

```
        1   1
      1   2   1
    1   3   3   1
  1   4   6   4   1
 1  5  10  10   5   1
1  6  15  20  15   6   1
```

heavier than the minimum. If the pattern is abcdefghij, the animal will be at the minimum weight. Similarly, animals with abCDEfghij or AbcdeFGhij will be equal to each other at 3 g above the minimum weight. If A or a and B or b, etc., are equally probable, the distribution of weights, in the population will then be identical to the patterns predictable from the statistical model of ten coins. If the number of heads showing in each toss is taken to represent the number of positive growth factors present, we can use an expansion of Pascal's triangle to the tenth level to obtain the relative frequency of heads. This will indicate the relative frequency of positive growth factors in the hypothetical growth problem, and in this way we can predict the expected relative frequency of individuals in each weight category. These results are summarized in Table 2-5 and shown graphically in Figure 2-2.

Figure 2-2 represents a type of curve commonly used in statistical work, the frequency distribution curve. The horizontal scale represents the variable under discussion, while the vertical scale represents either relative or absolute frequency. Note how this symmetrical distribution with the highest frequencies near the mean and vanishingly small frequencies at a distance from the mean conforms to our intuitive notion of relative frequency distributions as described in Chapter 1.

It is extremely important to emphasize how few factors need to be interacting in a random manner to produce a useful model of a frequency distribution. The preceding example was purely hypothetical;

Table 2-5

Wt. over Minimum, g	Proportion
0	0.001
1	0.010
2	0.044
3	0.117
4	0.205
5	0.246
6	0.205
7	0.117
8	0.044
9	0.010
10	0.001

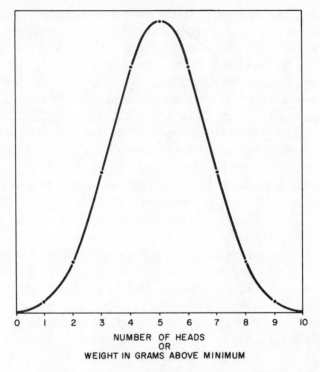

NUMBER OF HEADS
OR
WEIGHT IN GRAMS ABOVE MINIMUM

Figure 2-2. Hypothetical distribution of weights determined by ten equal factors.

therefore a comparison with figures from real life is of considerable interest. Figure 2-3 compares the frequency distribution of the heights of 100 men with the calculated frequency distribution of heads occurring among ten coins tossed 1024 times. Although probably many more than ten variables, genetic and otherwise, are determinants of height in these men, and although the data are derived from only 100 men, the comparison is fairly adequate.

Note that in this example we have used a discontinuous distribution (obtained from the theoretical consideration of tossed coins) as an approximation of the expected distribution of a continuous variable. This was accomplished by setting up 11 classes of weights each covering equal weight intervals on the horizontal scale. The line depicting the distribution of weights is called a *histogram*. Details of this method of processing the raw data and constructing histograms are given in Chapter 3.

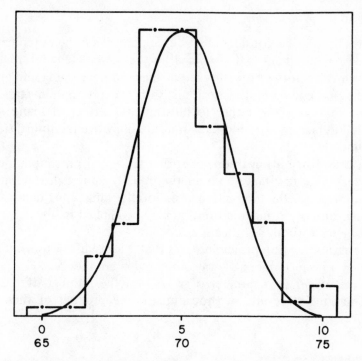

Figure 2-3. Comparison of theoretical and observed distributions of 100 measurements of height.

Although the histogram from real data is more or less approximated by the ten coin histogram, a general curve, which fits real data of continuous variables from a wide variety of sources, is the basis for most of the analytic methods of statistics. Although this curve was first described by De Moivre and later independently by Laplace, it is generally named for Gauss. Gauss produced an independent derivation of the curve related to errors in astronomical observations and developed many of its properties and uses.

In the hypothetical example used in Table 2-5, we assumed the existence of a relatively small number of factors producing a comparable number of discrete "steps" or classes of weight. Since continuous variables grow by an infinite number of minute steps, the *normal* or *Gaussian curve* may be expected to be related to this fact. Indeed, one of the several derivations for this curve is, in effect, the curve describing the relative frequency of heads when an infinitely large number of coins are tossed.

SUMMARY

A result may be considered to be random when the outcome of a trial or measurement is determined by the inconsistent interplay of many small factors in the absence of bias. In developing and applying the methods of statistical analysis, we use simple statistical models in two ways, either to illustrate the effect of many small factors interacting at random or to insure objective randomization in sampling.

Because distribution frequencies that are based on simple models with relatively few factors interacting rapidly approach the limiting form derived for the interaction of an infinite number of factors, this limiting form or normal distribution may be applied to a wide variety of problems without significant error.

The multiplication principle states that if a result or procedure can occur in a given number of ways, n_1, and if another result or procedure can occur in a number of ways, n_2, the number of possible combinations of results or procedures is the product of these two numbers, $n_1 n_2$.

PROBLEMS

1. A nickel, a dime, and a quarter are tossed at random. How many different and distinguishable combinations of heads and tails are possible?
2. Three dimes are tossed at random. How many readily distinguishable combinations are possible?
3. Why does Problem 2 have an answer different from Problem 1 when three coins are involved in both cases?
4. What would the values for the eighth line of Pascal's triangle, as shown in Table 2-4, be if the triangle were extended for two more levels?
5. a. Using the answer to Problem 4, calculate the probability of two heads and six tails appearing when eight coins are tossed.
   b. Using Table 2-4, calculate the probability of obtaining two heads and four tails when six coins are tossed.

3
RANDOM SAMPLING AND THE
DESCRIPTION OF DATA

Generally, statistical analysis in biology is concerned with inferences about a parent population. Previous discussions concerning the probable behavior of samples from a statistical model emphasized that predictions were valid only for the long run. Thus, large numbers of observations or large samples of data are preferable to small samples. Frequently, we are faced with the necessity of making the best inferences possible from small samples. However, these problems are discussed in later chapters. For the present we will be concerned with the theoretically simpler problem of the analysis of large samples. Where samples are sufficiently large, fairly extensive analytic inferences can be made on an intuitive basis with little or no recourse to the more sophisticated methods of statistics.

Assuming that we are dealing with large numbers of observations or measurements, we will have both the mechanical problem of making these more manageable by classifying and summarizing and also the problem of accumulating data that are representative of the parent population. In Chapter 2 the problem of random sampling was introduced in an elementary way. Of course, unless the sample represents the parent population properly, no inferences based on the sample are likely to be valid.

Whenever we draw a sample from a parent population, there is the assumption that no bias exists, which would favor one portion of the population while discriminating against another. We assume that the proportion of samples in a given category is very close to the proportion existing in the parent population and that nothing in our sampling technique introduces a bias.

Too often novices in statistical work fail to realize how easily bias creeps into a sampling technique. As a simple example of this bias, assume that we are given the task of measuring some characteristic of telephone users. We plan to study a sample of 100 subscribers to the service by selecting names at random from the telephone directory. Imagining the process in action, we can picture the observer closing his eyes, opening the book at random, and placing a pencil on the page before opening his eyes to see who is selected. Now it may or may not have any relevance to the nature of the survey, but it is more than likely that this apparently unbiased selection will favor those names in the middle of the book that appear near the middle of the page. Even if the observer is aware of this possibility, it will require a conscious and therefore nonrandom effort to avoid this bias.

We can, of course, use some variant of the primitive technique already described for selecting draftees. Numbered beans or cards representing individuals to be sampled can be randomized by being shaken in a hat or shuffled in one way or another. Frequently, this method is adequate, but for many purposes these simple mechanical randomizing methods are not only cumbersome but may superimpose a mechanical bias in some special situations. Thus, if the researcher uses numbered cards, he must be certain that the cards are completely uniform.

Since this is not an uncommon problem in the practical application of statistics, it should not be surprising to learn that a convenient and valid solution has been devised. With the aid of randomized computer systems, tables of random numbers have been generated. Table A-I in the appendix is an abbreviated example suitable for our purposes. To use this table, we assign a number to all members of the finite population to be sampled. If we have a group of about 100 items, there is no difficulty in numbering them from 1 to 100 in any order that is convenient. Where a very large number in the parent population is to be sampled, as in the problem of selecting 100 individuals at random from a telephone book, it would be impractical to do this. For this problem we would consider each name in the directory as being characterized by three numbers: one number would characterize the page number on which a name might appear; the second number would characterize the number of the column on the page, and the third number would characterize the number of the name in the column.

For the purposes of this illustration, assume that the directory being used has 875 pages with four columns of 100 names per column on each page. The first problem is to select 100 pages at random. Since the numbers in the table are already randomized, we may begin anywhere; therefore, in this instance we will start at the beginning of the table. We now copy three-digit figures in the order of their appearance until we have 100 figures less than 876 to represent the page numbers. Note that we take these figures as they appear and make no attempt to skip around in the table. The numbers are already randomized, and by taking them as they appear, we avoid unconscious number preferences, which might lead to bias. Referring to Table A-I then, our sequence would begin as follows: 100, (973), 253, 376, etc. Notice how the three digits are taken in horizontal sequence from their appearance in the two digit columns, and note further that the number 973 in parentheses above is to be discarded since it is larger than 875.

To select column numbers to go along with the 100 page numbers selected, we can begin with the second horizontal row and copy numbers in the sequence in which they appear until we have 100 numbers none of which is greater than four. Thus, we would have: 3, (7), (5), 4, 2 (0), 4, and so on. Again note that the numbers in parentheses are not applicable and must be discarded. The random selection process would now be completed by starting at some other point in the table and repeating the process for 100 three digit numbers equal to 100 or less. However, a large proportion of numbers in any of the sequences would have to be discarded, as we would be able to use only those three digit sequences in which the first digit is zero except for the number 100. To avoid this difficulty we assign the number 00 to the first name at the head of the column and number the remainder up to 99. Then we can use two digit sequences from the table and rapidly complete the task.

Once we have accumulated our large sample, we must still organize the data in an efficient and informative way. The sequence of procedures used to accomplish this may be illustrated with the data presented in Table 3-1. These represent serum calcium levels in 100 normal adults of both sexes. In order to obtain the average or arithmetic mean we could, if we wished, simply add all of the values and divide by 100. Even with only 100 figures this represents a fair amount of tedious work. With larger samples running into several

Table 3-1 Serium calcium levels in 100 normal adults

10.46	10.06	11.49	9.47	11.02
11.39	10.91	11.18	8.50	9.31
11.37	9.52	8.62	11.01	9.99
11.39	11.79	9.89	8.66	11.04
9.72	8.81	10.66	9.56	11.40
10.20	10.16	12.27	10.04	8.87
10.77	10.38	9.49	10.29	11.03
9.67	9.71	10.16	8.65	11.25
10.63	10.38	8.58	9.45	9.69
10.42	10.59	10.86	10.24	9.15
12.46	7.47	9.65	9.53	9.44
9.49	8.96	9.51	10.76	10.23
11.68	9.85	10.60	9.10	8.84
9.74	9.64	11.83	10.54	7.94
7.99	11.18	8.86	10.36	9.77
10.21	10.27	10.61	9.69	7.90
10.08	8.36	9.66	9.48	12.99
11.28	9.86	9.11	10.19	9.80
9.46	8.06	10.49	9.76	10.53
9.56	10.66	9.11	9.37	10.40

hundreds of items, this would be difficult even if we used an adding machine. Therefore the data are retabulated into groups. This process of grouping the data not only simplifies the computations of large amounts of data, but also illustrates certain simple inferences that may be made and which are useful in developing our ideas about statistical analysis.

In general, the number of subgroups should not be less than 10, while using more than 20 defeats the purpose of the process. Each group covers an equal interval, and the range covered by the sum of the intervals must equal or slightly exceed the range of the data.

Dividing the data into useful subgroups is generally easily accomplished by simple inspection, but it can be awkward and difficult unless a few simple rules and principles are kept in mind. In addition to the restriction that the number of classes shall be between 10 and 20, it is highly desirable to have the data fall easily into the appropriate groups. That is, we wish to avoid the situation in which a value falls on a boundary value between two classes. Arbitrary rules requiring the placement of such values into the next highest or lowest category tend to introduce some unnecessary bias in the analysis. To avoid this situation, we may think of the groups as compartments held together by links. The numerical value of each link is determined by the precision of the data. Thus, in the example under consideration, we note that none of the values contains figures more than two places to the right of the decimal point. Therefore we think of the range here as being made up of a chain of linked boxes in which each link represents the value 0.01.

In the trial formula below for estimating the size of the class interval, this link value for precision is represented by p. The highest value in the data being analyzed is M, while the lowest is m. Thus the range of the data is $(M - m)$, but for convenience the range to be covered by the groups may be extended so that the lowest value encompassed by the groups is $(m - y)$, while the highest is $(M + Y)$. Then if n is the number of groups to be formed and I is the size of the class interval:

$$I = \frac{(M - m) - (n - 1)p + (Y + y)}{n}$$

By inspecting the data and assuming that the data are to be divided into ten groups, we substitute the appropriate values in the formula to obtain:

$$I = \frac{(12.99 - 7.47) - 9(0.01) + (Y + y)}{10} = \frac{5.43 + (Y + y)}{10}$$

Since a convenient interval for I reasonably close to $5.43/10$ would be 0.55 or $5.5/10$, we let $(Y + y) = 0.07$. Rather than introduce unnecessary awkwardness, we arbitrarily let $y = 0.03$ and $Y = 0.04$ instead of making them equal. Using the values obtained in this way, we see that the lower limit of the first group will be 7.44, and since the class interval is 0.55, the upper limit of the group is 7.99.

We have already seen that the precision of the data establishes the "link" value as 0.01; therefore the next subgroup begins at 7.99 + 0.01 or 8.00 and extends to 8.55. The remaining classes or groups are defined by continuing this process.

Table 3-2

Class Boundaries		Tally	Class Frequency, f	Mid-mark, x	fx
7.44	7.99	///	4	7.715	30.860
8.00	8.55	///	3	8.275	24.825
8.56	9.11	LH1 LH1 //	12	8.835	106.020
9.12	9.67	LH1 LH1 LH1 ////	19	9.395	178.505
9.68	10.23	LH1 LH1 LH1 LH1 /	21	9.955	209.053
10.24	10.79	LH1 LH1 LH1 LH1	20	10.515	210.300
10.80	11.35	LH1 LH1	10	11.075	110.750
11.36	11.91	LH1 ///	8	11.635	93.080
11.92	12.47	//	2	12.195	24.390
12.48	13.03	/	1	12.755	12.755

Table 3-2 shows the groups that are formed and how these are used to make a tally of the values that fall into each of the groups. In the remaining calculations, the class midmarks represent all of the values in each group. As indicated in the table, the sum of the products obtained by multiplying each of the class midmarks, X, by the class frequency, f, is taken as being equal to the sum of the individual values. Although this value may differ slightly from the sum that might be obtained by adding each of the ungrouped values, this difference becomes negligible when divided by the total number of items in finally obtaining the mean. Of course, this does not change the percentage error incurred, but even this is apt to be either negligible or of no consequence in subsequent calculations.

More important, at this point, than the computational details is the overall appearance of the tally marks. We can see that this forms a crude picture of a normal frequency distribution turned on its side. Since a diagonal mark is made in each group of five in the tallying process, the varying lengths of the tally are not quite proportional to the total frequencies in the groups. Figure 3-1 represents the tallying process by the distribution of flat uniform counters into boxes representing the groups.

In this diagram the heights of the piles of counters are directly proportional to the frequencies in each of the categories designated on the horizontal axis by the midmark values for each group. If we

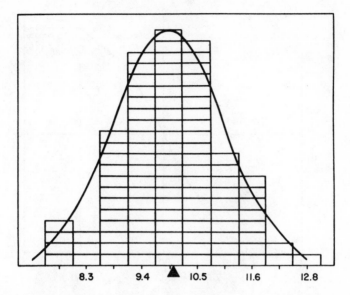

Figure 3-1. Relationship of mean to balance point of a distribution of tallies of equal weight.

had actually placed tally counters of uniform dimensions into a container as represented, the center of gravity of the distribution would be located above the value for the mean as indicated by the dark triangular fulcrum on which the horizontal axis is balanced.

Since the sample is relatively large, we intuitively believe that if we were to take another equally large sample, the results would be quite similar with a new mean quite close to the mean of the first sample. Although we would not expect the frequencies in each category to be exactly the same, we would expect a reasonable degree of similarity. In short, we feel reasonably secure that a large sample of data yields a fair representation of the parent population.

To exploit these notions a little further, we summarize and abstract our information by forming a cumulative frequency distribution curve as shown in Figure 3-2. More will be said about such curves in later chapters. This curve is formed by accumulating the percent of the sample that is included below the values shown on the horizontal scale of the graph. Thus, beginning at 7.44 (the lower limit of our first category), we see that 0% of our sample population is as low or lower than this value. We do not necessarily conclude that no example of a lower value exists in the parent population, but it is not

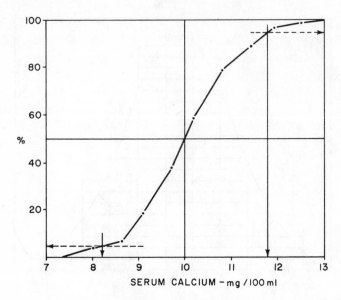

Figure 3-2. Cumulative frequency distribution curve for same data as shown in Figure 3-1.

unreasonable to infer that the probability would be negligibly small of finding a lower value for a randomly selected individual. Although empirical curves obtained in this way are not always so well-behaved, in this instance the curve reaches the 50% level at almost exactly the level of the mean.

The broken lines on the diagram are drawn horizontally at the 5% and 95% levels. Seeing where these intersect the curve, we can predict with some confidence that only about 5% of the normal population is likely to show a serum calcium level below 8.21 mg/100 ml or above 11.8 mg/100 ml. Thus, if we choose to regard levels outside this range (i.e., 8.21 – 11.8 mg/100 ml) as abnormal, we will wrongly investigate patients for additional evidence of disordered calcium metabolism only about 10% of the time.

Returning again to Figure 3-1, we see that the essential features of the distribution are adequately represented by the silhouette of the piled up tally counters. Thus, on future occasions, a picture of a pile of counters will not be necessary. Instead, we can mark off the boundaries of our groups on the horizontal scale and represent the overall silhouette by making off rectangular columns above these values. The resulting diagram is called a frequency polygon, or more properly, a *histogram*. Superimposed on the histogram is a normal distribution curve. Generally, the larger the sample, the more closely does the histogram correspond to a normal distribution curve.

From these diagrams we see that the *mode* or most "popular" value is in the same column as the mean. The crest of the normal distribution curve (i.e., the mode) coincides exactly with the mean. As we noticed above, the *median* value, the value which has an equal number of individual values above and below it (i.e., the 50% level), also coincides with the mean.

By examining the histogram further, we realize that the cumulative frequencies, as shown by the curve in Figure 3-2, represent the corresponding areas of the histogram below the values being considered. Thus, the term, *median,* refers to the demarcation separating one half of the area from the remaining half; the *mean,* on the other hand, refers to the balance point below the center of gravity of the structure, while the *mode* refers to the most common value. In a normal distribution the mean, median, and mode coincide. Other kinds of distributions, in which these values do not coincide, will be considered later in this book.

SUMMARY

It may be taken as axiomatic that larger samples are likely to represent the distribution of values in a parent population more accurately than small samples. This is true only if the sampling of the parent population has been truly random and free of bias. Randomization of samples may be achieved with various mechanical devices, but the easiest way to accomplish this is to use tables of random numbers.

Large samples of data are most efficiently handled by grouping the data into no less than 10 and preferably not more than 20 evenly spaced categories or classes and tallying the frequency in each class. A frequency polygon showing the relative frequency of individuals in each class is called a histogram. Generally, the larger the sample is, the more closely does the histogram conform to the outline of the normal distribution curve. The mean of the values represents the balance point below the center of gravity of the polygon; the median divides the area of the polygon representing the distribution of the sample population in half, while the mode represents the category with the highest frequency. The median is that value which has as many individuals below it as above. In a normal distribution the mean, median, and mode coincide.

PROBLEMS

1. To make a random selection the entries in Table 3-1 were numbered starting at the top of each column. The first entry at the top of the column on the left, 10.46, is numbered 01; the last entry at the bottom of the last column on the right, 10.40, is numbered 00.

   Starting at the top of Table A-1 select 11 numbers for a random selection from the columns numbered above.
2. For these 11 randomly selected numbers find the median and calculate the mean.

4
THE NORMAL DISTRIBUTION CURVE

At the end of the last chapter the subject of the Gaussian or normal curve was introduced. This curve, which is the basis of most of the applications of statistics, is also referred to as the "error law." The use of the term *normal* in describing this distribution is quite popular, but as will be seen, there are other useful distributions that are quite different but cannot in any way be considered abnormal. The term "error law" has historical interest, since the classic derivation by Gauss of the equation for the normal curve was based on an analysis of the distribution of errors or deviations around the mean of precise physical measurements.

Commonly in an analysis of a set of data we make the working assumption that the data being sampled would fit the theoretical distribution of the normal curve very closely. Of course, real data rarely coincide exactly with a theoretical curve, but the consequences of small deviations of real from theoretic values are usually negligible. As has been implied, the reason for the closeness of the approximation of the theoretical normal curve to real data lies in the random interaction of many small and variable factors. As demonstrated in Chapter 2, the distribution of results arising from the random interaction of as few as ten equal factors follows a curve not too different from the normal curve. In experimental data or sets of observations the interacting factors need not be assumed to be equal. These factors need only be small and numerous. Their random nature will generally insure that unequal factors will be operative as often in one direction as in the other and in any long run will probably be balanced by a comparable number of factors of similar magnitude operating in the opposite way.

This does not necessarily imply that we may always use the normal distribution in the analysis of data merely because we know that the factors are many and small. Occasionally, the particular measurement or observation being used is a nonlinear function of some other variable which is normally distributed. The following example should serve to illustrate this situation. Suppose, for example, we are measuring the *diameters* of branches of the bronchial tree as they appear in a cross section of the lungs. Since the factors determining the diameters in a given cross section are certainly numerous and small, we would naturally assume that the data would follow a normal distribution. Suppose now that a researcher with similar interests observes the same or similar material and decides to accumulate the data on the cross sectional *area* of the observed bronchial branches. Using the same reasoning, he would be equally justified in assuming that his data would fit a normal distribution. While either one might be right, and while it is possible that both are wrong, it would be mathematically impossible for both to be right. The cross sectional *diameter* of each tube is directly proportional to the radius, but the cross sectional *area* is proportional to the *square* of the radius. Thus, if a graph of the distribution of the diameters were symmetrical, a graph of the areas would have to be nonsymmetrical or skewed. On the other hand, if the distribution of areas were symmetrical, the distribution of diameters would be skewed in the opposite direction.

It is particularly important for biological workers to understand and to recognize these forms of distortion in the normal distribution. Many physiological measurements are related in some way to the size of the subject. Size in this instance may refer to the height, to the surface area, or to the weight, which is more or less closely related to the volume. As an example of how data may be distorted by selecting the wrong units as a basis for analysis of a distribution, think of an organism as being roughly equivalent to a cylinder. This oversimplification may not apply too closely to man or to most other organisms, but for eels or earthworms the abstration is not too drastic. In any event the principles to be illustrated by this abstraction hold true as a generality surprisingly well.

Figure 4-1 illustrates three cylinders representing three animals of the same species. Since they are of the same species, we may safely assume that in these specimens the ratio of the height to the radius is

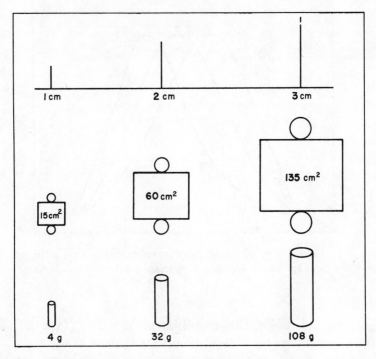

Figure 4-1. Comparison of dimensions in a species of cylindrical organism.

practically constant. The first row illustrates the relative heights of these animals; the second row illustrates the relative surface areas (the circles correspond to that part of the surface area of a cylinder that would be the lid and bottom); the lower row of figures shows the relative volumes, and because of the limitations of plane figures in showing three dimensional relationships, the relative weights are given below.

Figure 4-2 compares the distributions based on height, surface area, and volume. To understand the reason for the distortions of the normal shape as shown for the surface area distribution, we may consider each of these measurements as a function of the radius. As indicated above, the height is more or less proportional to the radius within a given species. Thus, we may express the height, h, in terms of the radius, r, as $h = Kr$. The surface area, a, is related to height and radius as shown in the following formula: $a = 2\pi r^2 + 2\pi rh$. However, if we substitute the relationship given above between the

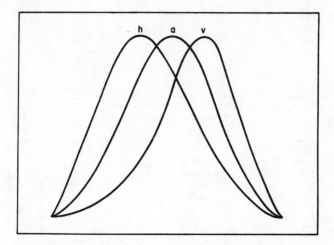

Figure 4-2. Comparison of distribution curves resulting from different dimensions of the same species (h = height, a = surface area, and v = volume or weight).

height and the radius, this simplifies to: $a = 2\pi r^2 (1 + K)$. Thus, surface area is directly proportional to the square of the radius. The volume, v, of the cylinder is related to the height and radius as follows: $v = 2\pi r^2 h$; through substitution as in the previous example, this becomes $v = 2\pi r^3 K$. In other words, the volume, and indirectly the weight, are related to the *cube* of the radius.

Two further sources of distortion are commonly encountered in physiological or biochemical studies and in pharmacological investigations. In the former group, typically in the estimation of enzyme activity, careful distinctions must be made between the rate of a reaction and the time required for the reaction to reach a predetermined end point. If the time required to reach an end point is measured as t, the reaction rate, which is more likely to reflect the relative activity of an enzyme, is proportional to the reciprocal of time, $1/t$. Thus, if the enzyme concentrations in a given situation are normally distributed, the distribution curve based on end point times will be skewed. The mean obtained from the *reciprocals* of the observed time values is the *harmonic mean*.

In pharmacological studies relating the effect of some drug, hormone, or other active material to the dose, another variety of factors must be considered. In general, the rate of removal or inactivation of any of these active compounds from the blood or

Table 4-1

Dose	Duration, hr
1 unit	1
2 units	2
4 units	3
8 units	4

body fluids is directly proportional to their concentration. Because of such complicating factors as the rate of equilibration between the various fluid compartments of the body, there are often deviations from this simple rule, but it still provides a satisfactory approximation for most statistical analyses. Thus, we can state that the time required for sufficient elimination of a drug to reduce its concentration in half is more or less constant. If the initial concentration is proportional to the dose, we can assume that the dose must be doubled for each standard increase in effect or duration of action. Table 4-1 shows the relationship of dose to duration for a hypothetical drug. Because of the above considerations, it is commonly found in the statistical analysis of pharmacologic data that normal distributions are obtained if the *logarithm* of the dose instead of the value for the dose itself is used as the scale for the horizontal axis of the distribution. The antilogarithm of the arithmetic mean of these logarithms is the *geometric mean*. Thus, the geometric mean of the doses listed in Table 4-1 is 2.828 (i.e., the antilog of 0.4515).

In addition to the mathematical causes of deviation from normal distribution, we must be alert to detect biasing factors which operate in one direction only. For example, suppose we were studying the distribution of intelligence measurements. Just as with the diet requirement example in Chapter 1, we would intuitively guess that most of the population would be evenly distributed close to the average. We would also expect to find about as many geniuses as idiots if intelligence were normally distributed. However, there are accidents and diseases of one kind or another that can cause a decrease in intelligence, and we know of no disease, drug, or accident that can increase intelligence. Thus, although the factors determining intelligence are many and small, the existence of biasing factors can produce a distortion in the final distribution.

These precautionary comments should not be taken as a warning against the assumption of normal distribution for data presented for analysis. Indeed, as will soon be indicated, many statistical methods assuming normality of distribution can be applied to collections of data that deviate markedly from normal. Of course, the statistical inferences are more valid and accurate if the units of such data are modified so as to produce a normal distribution curve. In any event, inferences regarding extremes even of normal distributions should be made with caution.

The Gaussian or normal curve is described by the equation:

$$y = \frac{1}{\sigma\sqrt{2\pi}}e^{-(x-\mu)^2/2\sigma^2} \tag{1}$$

To the nonmathematical student this equation will certainly appear formidable, but since it is not used directly in biological statistics, there is no need for undue apprehension. There are a few significant features of this equation, which can be appreciated by those who are relatively naive mathematically. In this equation y represents the relative frequency of some variable quantity, x. The values for π and e are constant; π is the familiar ratio of the circumference to the diameter of any circle, 3.1416; e is the base of the Naperian or natural logarithms. If this last item is totally unfamiliar to the student, he need not be concerned at this point, but may accept e as a constant in the same way that the value of π is accepted.

The important features to be noted are the two parameters (i.e., numerical characteristics) μ and σ. The first, μ, is the arithmetic mean; the second, σ, is called the standard deviation. As we shall see, the standard deviation is a measure of the spread of the data about the mean. Since the values for π and e are fixed forever and are constant, the entire curve is completely defined or characterized by the two parameters, the mean and standard deviation.

Referring now to the normal curve defined by Eq. 1, we may take note of certain characteristic features as shown in Figures 4-3 and 4-4. It is bell shaped, and, as previously noted, symmetrical. The median and mode coincide with the mean, μ. If we scan the curve in either direction away from the center, we find that it is initially convex upward, but soon becomes concave. There is a point of transition from convex to concave curvature called an inflection

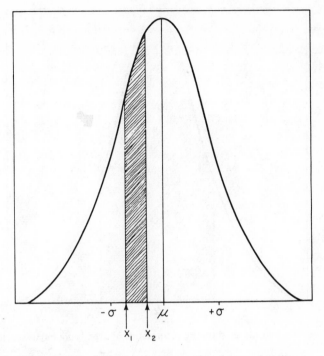

Figure 4-3. Normal curve showing area proportional to probability.

point. The distance of this inflection point horizontally from the mean is equal to the standard deviation, σ.

Another feature of interest is the way in which the curve tapers out to infinity in either direction from the mean before touching the horizontal axis. This may not be obvious from the diagram, but it is clearly implied by the equation for the curve. This also implies that there is some theoretical probability for any value of the variable x. Some students are disturbed by what must be an obvious fiction. Thus, if we were showing a theoretical distribution of life expectancies, it would be meaningless to compute a finite probability that a person would live to be 1000 years old, or even more fantastic, the probability for a negative value for life expectancy. However, two points must be kept in mind. First, the finite probabilities become so vanishingly small at these relatively extreme distances from the mean as to be indistinguishable from zero. Second, and perhaps more important, the normal curve may be a very close approximation to real distributions, but it is not necessarily exact, and, as has already been noted,

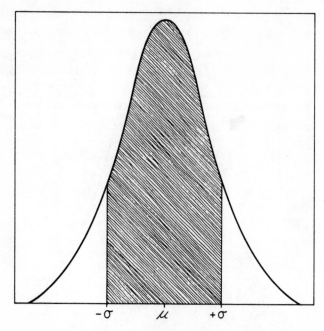

Figure 4-4. Normal curve showing proportion of area lying between $(\mu + \sigma)$ and $(\mu - \sigma)$.

considerable caution must be used in interpreting calculations of finite probabilities for extreme values of the variable.

Before getting down to the practical problems of relating a set of data to the normal curve, one should consider briefly, as a preview of potential applications, some useful and interesting features of the curve. First, the area under the curve is taken as unity or 1.0. That is, the sum of all the possible relative frequencies (i.e., probabilities) represents certainty. Thus, the area lying under the curve between the values x_1 and x_2 shown as the shaded area in Figure 4-3, represents a fraction of the total area proportional to its probability. This fraction represents the probability of obtaining by chance a sample value lying between the values x_1 and x_2. Similarly, note that as a consequence of the curve's symmetry, half the area under the curve lies above the mean while half lies below. This implies that there is a probability of 1/2 that any value randomly selected will be greater than the mean. There is an equal probability, 1/2, that it will lie below the mean.

The shaded area shown in Figure 4-4 is the area lying under the curve between the inflection points. As has already been indicated, the distance of these inflection points from the mean defines the value for the standard deviation, that is, the value of the parameter, σ, in Eq. (1). This area constitutes about 2/3 of the total area, or more exactly 68.3%. The relevance and utility of these facts will become clear after the process of estimating these parameters has been shown.

The method of calculating the average or arithmetic mean is generally sufficiently well known to require little if any discussion. For this reason it may serve to introduce the abstract algebraic notations commonly used in statistics. As indicated above, x usually stands for the value of some variable. Using this symbol, we indicate the existence of some series of measurements of the variable x with a subscript indicating the position of the value in the sequence of measurements as follows:

$$x_1, x_2, x_3, x_4, \text{etc.}$$

Generally, we indicate the existence of a number of measurements, n, merely by indicating the first two or three values followed by three dots and ending in x_n, thus:

$$x_1, x_2, x_3, \cdots x_n \tag{2}$$

The mean of these values is generally indicated as $\bar{x}$ (read as x bar). The first step in calculating the mean is addition of all the values of x. The symbol for this sum is Σx. Thus, $\Sigma x = x_1 + x_2 + x_3 + \cdots + x_n$. Dividing this sum by the total number of measurements yields the mean. Thus, we write $\bar{x} = \Sigma x / n$.

Usually, the novice feels that the average deviation from the mean would be an adequate measure of the dispersion about the mean. There are two difficulties with this approach. First, if due consideration is given to the sign of each deviation, the positive deviations and the negative deviations will add up to zero. Second, if an addition of deviations is made ignoring the sign, a fair index of spread for an individual case will be obtained, but statistically, this would be a "blind alley." We would have an index of spread, but nothing further of any statistical use could be made of the value.

For statistical purposes the scatter about the mean is best expressed as the *standard deviation*. This is calculated from the data as the square root of the mean of the squared deviations. First we obtain the deviations of each of the measurements from the mean. This step is indicated in the symbolic notation of statistics as:

$$(\bar{x} - x_1), (\bar{x} - x_2), \cdots (\bar{x} - x_n) \tag{3}$$

About half of these will be negative values, but in the next step each value is squared, so that they all become positive. Thus, we have:

$$(\bar{x} - x_1)^2, (\bar{x} - x_2)^2, \cdots (\bar{x} - x_n)^2 \tag{4}$$

The mean of these squared values yields a new value known as the variance, V, which will be put to additional use in a later chapter: Thus:

$$V = \frac{\Sigma(\bar{x} - x)^2}{n} \tag{5}$$

The estimated standard deviation, s, is the square root of the variance:

$$s = \sqrt{V} = \sqrt{\Sigma(\bar{x} - x)^2 / n} \tag{6}$$

When samples with more than about 30 measurements are analyzed, Eq. (6) may be used without modification. For smaller samples this formula tends to underestimate the standard deviation. This is a result of the tendency of a small sample to underrepresent the less common wider variations from the mean. It can be shown that the best formula for standard deviation substitutes $(n - 1)$ for n in Eq. (6) as follows:

$$s = \sqrt{\Sigma(\bar{x} - x)^2 / (n - 1)} \tag{7}$$

While the effect of this correction is important when the samples are relatively small, the magnitude of the effect of the correction diminishes rapidly as the sample size is increased. When, for example, $n = 30$, the magnitude of the correction is around 3% of the variance, but when the calculation of the standard deviation is completed by taking the square root, the effect of this correction diminishes to less than 2%.

If the sample is large enough and the distribution is normal, the entire distribution is characterized in terms of the two parameters, mean and standard deviation. Recall that Eq. (1) for the normal distribution curve is uniquely determined by these parameters. With this equation or with the corresponding curve, we could state what proportion of the population sampled would fall between any two values, x_1 and x_2. We could do this crudely by measuring the fraction of the total area thus delineated under the curve. Using the methods of integral calculus, we could make an accurate calculation, but this is both laborious and one of the more difficult mathematical processes. In practice this value is easily obtained with a table already worked out. The basis for being able to use a single table for the infinite variety of distributions, which may be encountered, is as follows.

Figure 4-5 shows two identical normal distributions located side by side on the same horizontal axis. We can see that there is no change in shape or area under the curve when the position of the mean is shifted on the scale. If we shift the scale so that the mean is at zero, this is comparable to preparing a histogram by replacing the original values of the variate by their deviations from the mean. In other words, we are replacing $x_1, x_2, x_3, \cdots x_n$ by the sequence of deviations, $(\overline{x} - x_1), (\overline{x} - x_2), (\overline{x} - x_3), \cdots (\overline{x} - x_n)$.

Figure 4-6 shows two normal distribution curves that have been shifted laterally so that the means of both curves are at zero. Although one curve is broader, having a larger standard deviation

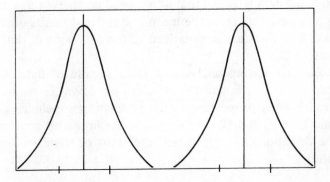

Figure 4-5. Changing the location of the mean by addition or subtraction does not change the shape of the curve or the value of the standard deviation.

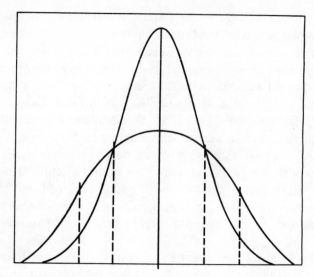

Figure 4-6. Two normal distributions with their means at zero, but with differing standard deviations.

than the other, in both curves the same proportion of their areas (i.e., about 2/3) lies between their inflection points at +s and −s (indicated by the vertical dotted lines). This suggests that we could refer both of these curves, or any curve of a normal distribution, to a standard curve with its mean at zero and its inflection points at +1 and −1. In other words, the horizontal unit of the standardized reference curve is one standard deviation. The areas included between any two paired deviations from the mean on this curve have been calculated and tabulated, so that all we need do to use this table is to convert any deviation from the mean in our data to units of standard deviation. The steps are summarized in the following outline of the procedure.

Suppose that having analyzed a large sample of data, we have calculated the mean, $\bar{x}$, and the standard deviation, s. We now wish to determine what proportion of the population is included between the values $(\bar{x} + x)$ and $(\bar{x} - x)$ shown as the unshaded area in Figure 4-7. The deviations are converted into units of standard deviation, U, as follows:

$$U = \pm \frac{\bar{x} - x}{s} \qquad (8)$$

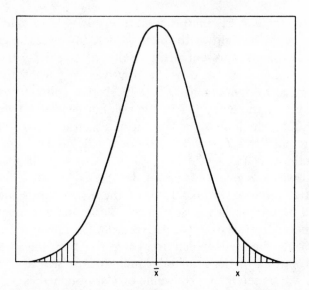

Figure 4-7. Shaded areas show proportion of a normally distributed population deviating from the mean by more than $\pm(\bar{x} - x)$.

The proportionate areas corresponding to U are given in Table A-2I and II. In this table the values given the proportion of the total area under the normal distribution curve lying in the shaded portions corresponding to various values of U. Since the total area is equal to unity, the unshaded area is calculated as $(1 - p)$.

As a concrete example, imagine that we are conducting an experiment using spherical seeds, and for the purpose of the study, we want all of the seeds used in the experiment to deviate from the mean diameter by no more than 5% of the mean. Previous studies on these seeds have revealed that the mean diameter is 10 mm with a standard deviation of 2 mm. We separate the seeds that are too large by passing the seeds through a sieve with holes 10.5 mm in diameter (i.e., 10 mm + 5% of 10 mm). The seeds to be used are separated by passing the remainder through a sieve with holes 9.5 mm in diameter (10 mm – 5% of 10 mm). From the available information we are asked to determine how many useful specimens we will have if we start with 1000 seeds.

First we calculate the value for U:

$$U = \frac{(10 - 9.5)}{2} = 0.25 \qquad (9)$$

Turning to Table A-2I, in the column headed U, we find 0.2. The next column to the right, with the heading 0.00, gives the proportion of the shaded area represented by a value of $U = 0.2 + 0.00$ or $U = 0.20$. The value here is 0.8415. However, we want the area or proportion corresponding to $U = 0.25$. In the column headed 0.05 (at the level of the entry 0.2 in the column headed U) we find the entry 0.8026 which may be appropriately rounded off to 0.803. This is the entry for $U = 0.2 + 0.05$ or $U = 0.25$. The proportion of seeds retained is $(1 - 0.803)$ or 0.197 which multiplied by the original 1000 seeds gives the answer, 197 seeds.

Similarly, suppose we take 1000 of these same seeds and ask how many will be retained by a sieve with holes 6 mm in diameter. Note how the nature of this example differs from the previous one. In this instance, only the seeds which are 4 mm *less* than the mean will be discarded, but all of the remainder including those 4 mm larger than the mean will be retained. The value of U is obtained as:

$$U = \frac{(10 - 6)}{2} = 2 \tag{10}$$

In the language of statistics the table is "double-tailed," but for this particular example we require a "single-tailed" value. That is, the proportionate area, p, represents both of the shaded areas represented at the ends of the distribution curve in Figure 4-7. These areas are equal, and since we only require the area of the shaded tail on the left, we take half of the tabulated value. In this instance, the table shows that for $U = 2$, the proportion of the population included under both shaded areas is 0.045. However, only half of this value represents the seeds that have diameters less than 6 mm and will fall through the sieve. Thus, 0.0225 (1000) represents the number of seeds we could expect to exclude. This would not exceed 23, and therefore we would expect the sieve to retain 977 seeds.

These examples and the problems at the end of the chapter illustrate some of the immediate direct applications of the mean and standard deviation. They also reveal the necessity for determining by the nature of the problem whether a single- or double-tailed value is to be used. Tables of the proportionate areas under the normal curve are fairly widely available, but the value of the deviation in units of standard deviation are not always referred to as U; some tables present single-tail values, while in other the tabulated values refer to the area corresponding to the unshaded portion of the curve shown in Figure 4-7. For convenience in later discussions, we shall adopt the convention of letting U stand for the double-tail values and Z for single-tail values. In the example above, if we had looked up the entry for Z in Table A-2II, we would have found the proportion corresponding to Z = 2.0 to be 0.0228.

SUMMARY

When many small factors acting at random determine the final value of a variate, the frequency distribution of the variate will usually follow a normal distribution. Although real data rarely fit the normal distribution precisely, predictions based on the assumption that they do are generally valid. Even when distributions show obvious departures from normality, the errors made by using this assumption are usually small. Nevertheless, the accuracy of the predictions are improved when the curve closely approaches the normal shape.

Therefore, appropriate changes in the dimensions used as a basis for observed measurements are frequently desirable. In biological material the observed measurement may be a root or power of some fundamental dimension of the organism which is distributed normally. These possibilities are illustrated by the relationships between height, surface area, and volume or weight. Similarly, reaction times are often more apt to be normally distributed if converted to reaction rates, while in dosage problems the logarithm of the dose producing an effect is usually normally distributed. The harmonic mean of a series of values is the reciprocal of the arithmetic mean of their reciprocals. The geometric mean of a series of values is the antilogarithm of the arithmetic mean of their logarithms.

FORMULAS

Mean:
$$\bar{x} = \Sigma x / n$$

Variance:
$$V = \Sigma (\bar{x} - x)^2 / (n - 1)$$

Stand Deviation:
$$s = \sqrt{V} = \sqrt{\Sigma(\overline{x} - x)^2/(n-1)}$$

Relative Deviation:
$$U \text{ or } Z = \pm(\overline{x} - x)/\sigma$$

PROBLEMS

1. Calculate the mean and standard deviation for the first ten values in the first column of Table 3-1 after rounding off to eliminate the second figure to the right of the decimal point.
2. Assuming that the mean and standard deviation found in Problem 1 accurately represent the general population, what are the limits about the mean that will include 95% of the normal population? What are the limits that will include 99%.
3. What proportion of the population will be as high or higher than 11.0?
4. What is the probability that five normal subjects selected at random from the above population will all show serum calcium levels between 9 and 11 mg/100 ml?
5. A chemist evaluating a new analytical technique reports that 50% of the time his assay is within 0.1% of the correct value. What is the probability that a given assay might underestimate by as much as 0.05%?

5
SAMPLES AND THE UNIVERSE OF DISCOURSE

Up to this point, the discussion has generally implied that the estimates of the mean and standard deviation were accurately known because of the large sample sizes. However, even with fairly large samples, we know that if we were to repeat the entire experiment or series of observations with an equally large sample, the new mean and standard deviation would not match the first set of parameters exactly. The alert reader probably will have noticed by now that we have carefully avoided specifying how large a number of observations is considered a large sample.

Before considering this problem more definitely, we need to consider a few of the terms we have used with some clarity and precision. First, there are various ways in which data are generated. Data may arise from certain mechanical or physicochemical measurements or from samplings of a defined group of phenomena or beings. As examples of the first category, we may take the measurement of the speed of sound in a certain medium or, more in keeping with our interests, a series of measurements of the amount of calcium in a quantity of bone ash. Because of limitations in the precision of our measuring techniques, we know better than to expect repeated measurements to check exactly. Part of the difficulty resides in the refined accuracy that we may desire. If we measure the diameter of a cylinder in a metabolism machine by using a simple meter scale and are satisfied with 28 mm as an answer, repeating the measurement a million times will give the same answer a million times. There is no statistical problem here. On the other hand, if I measure very carefully with a micrometer and obtain the answer 27.996 mm, a second measurement may read 27.998. If for some specific reason, this order of accuracy is needed, I will make additional readings. Theoretically, I could make an infinite number of readings. The same could be done with the examples above to generate *populations* of measurements of the speed of sound or populations of estimates of calcium content. In this instance, the statistics of physicochemical measurements borrows a term from the demographers.

Without specifically saying so, we act on the implied assumption that there is a definite "true value" and that the mean of an infinite number of measurements would coincide exactly with this "true value." Such assumptions create philosophic difficulties, but in the situations in which we operate as biologists these are without relevance.

The measurements in the second category can be exemplified by the mean normal hemoglobin concentration or the average diameter of redwood trees. In principle, since there is a finite number of normal people on this earth, we could measure the hemoglobin concentration in all of them and arrive at an average or arithmetic mean. Since this is impractical, we can obtain an *estimate* of this mean. We tend to assume that the more people we measure the closer will our estimate approach the universal mean. In the physical measurements cited above, this assumption might hold and be borne out by experience. In making a physical measurement, the shifts in the value of the mean would either creep in one direction to a limiting value or show only restricted oscillations about a limiting central value. By contrast, when the size of a sample is increased, relatively abrupt shifts might occur unless special randomizing procedures are used. These shifts would reflect the existence of subpopulations in the overall population. This would be analogous to finding one population of physical measurements for one cylinder and a slightly different and distinctly separate population of measurements on another cylinder. In order to specify which set of a theoretically infinite number of measurements are being considered statisticians use the expression *universe of discourse.*

We recognize that the parameters of any universe of discourse can never be known exactly. However close our estimates may come to "true" values, the mean and standard deviations of any universe of discourse represent an ideal but generally unknowable value. We distinguish between the ideal value and the measured estimate in our algebraic notation by using Greek letters for the ideal values and Roman letters for estimates. Thus, the ideal mean for a universe is usually symbolized by the letter μ and the corresponding standard deviation by σ. The measured estimates use $\bar{x}$ or m for the mean and s for the standard deviation.

Although we may be unable to attain the ideal mean of an infinite number of measurements, we can arrive at a satisfactory evaluation of the magnitude of error in an estimate by considering the effect of sample size on the mean. In Figure 5-1 the curve for $n = 1$ represents the normal distribution curve of a population whose mean is μ and standard deviation is σ. Suppose that in estimating the characteristics of a population, instead of making individual measurements we make the measurements in pairs and treat the means of the two measurements as a variate. Intuitively, we can see that this derived population of means will have a distinctly narrower scatter, but obviously the mean of these pairs will be identical to the universe mean of the parent population. This narrower, steeper distribution is shown superimposed on the parent population in Figure 5-1.

If we repeat this process using groups of four in each sample and plotting the means of these subgroups, the mean of these sample means will be identical to the universe mean. This curve, however, is even narrower and steeper than the curve for the means of sample size of two. Thus, as we increase the size of the samples making up the subgroups the means remain unchanged, but the standard deviations

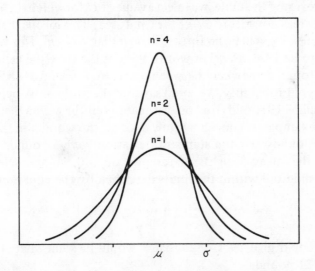

Figure 5-1 Effect of sample size on distribution of sample means.

of the distributions of sample means diminishes. It can be shown that if σ is the standard deviation of a universe, the standard deviation of means for samples of n variates is $\sigma/\sqrt{n}$. This value is usually called the *standard error of the mean*.

Each of the increasingly narrow distributions described above is still a normal distribution. It follows then that we can utilize the tabulated values of proportionate area expressed as a function of the standardized deviate, U. Thus, if we have a mean, $\bar{x}$, and estimated standard deviation, s, we can write:

$$U = \pm \frac{(\mu - \bar{x})}{s/\sqrt{n}} \tag{1}$$

Note that in this expression s is used as if it were identical or at least not appreciably different from σ. By simple algebraic transformation, we may rewrite this as:

$$\bar{x} = \mu \pm Us/\sqrt{n} \tag{2}$$

Written this way the equation states that the estimate of the mean using a sample size, n, lies within a certain distance of the universe mean, μ, if we can select an appropriate value for U.

Suppose, using the table, we select a value of U for which the excluded or shaded areas constitute 5% of the total area. This means that 95% of the estimates, $\bar{x}$'s, will lie no further from μ than $Us/\sqrt{n}$. This value will be found to be 1.96. In other words, 95% of the time the estimate, $\bar{x}$, would be found somewhere between $\mu + 1.96 s/\sqrt{n}$ and $\mu - 1.96 s/\sqrt{n}$; or in terms of probability, we could say that the probability being 0.95, there would be 19:1 odds that our estimate would lie within these limits.

If, for example, the mean weight, μ, of all the women in the world were 120 pounds and the standard deviation were 5 pounds, 95% of the time the mean weights of samples of 100 women selected at random would fall within the limits described by the equation:

$$\bar{x} = 120 \pm 1.96(5/\sqrt{100}) \tag{3}$$

Thus, we could give 19:1 odds that $\bar{x}$ would lie somewhere between 119 and 121 pounds.

The difficulty at this point is that we do not know that μ is 120. Suppose, however, that our sample mean, $\bar{x}$ on a sample of 100 women turned out to be 119 pounds. Since $s = 5$, we can see that this is consistent with all that has been said so far. However, if μ were really 118 instead of 120, the observed value for $\bar{x}$, 119, could happen with the same degree of probability. Thus:

$$\bar{x} = \mu + Us/\sqrt{n}$$

$$119 \cong 118 + 1.96(5/\sqrt{100}) \tag{4}$$

We see then that if there were two populations such that in one $\mu_1 = 120$, and in the other $\mu_2 = 118$, as long as both had a standard deviation $s = 5$, 119 could be obtained as a sample mean from either population with the same degree of probability. It follows from this that if $\bar{x}$ is 119, it could have come from a population with a universe mean of 118, but we have no way of knowing which one is the parent population. However, we can reverse the reasoning and say that if $\bar{x} = 119$ and $s = 5$, the odds are 19:1 that μ lies between 118 and 120. Thus, we can write:

$$\mu = \bar{x} \pm 1.96s/\sqrt{n} \tag{5}$$

Now 19:1 odds are pretty good, but why stop there? We can, of course, select whatever odds we desire, but in most scientific work, particularly in the biological sciences, 95% probability is conventionally accepted as adequate assurance for two reasons. First, there is little gain in assurance when stringent demands are made. If the limits are extended from 1.96 or $2s$ on either side of the mean to $2.58s$, the probability goes from 95% up to 99%. Thus, for a 30% increase in range around the mean, we gain a mere 4% in increased assurance. Some occasions may require this assurance, but these are rare. Furthermore, demanding more than 95% assurance is apt to lead to a disproportionate rejection of true hypotheses with little compensatory gain in rejection of false hypotheses. Conventionally then, we speak of the limits around the mean, $\pm 1.96s/\sqrt{n}$, as the 95% confidence limits or *fiducial limits*.

To make the meaning and usefulness of this maneuver more apparent, a real compilation of data will be considered. In a tabulation of normal values in clinical medicine the mean hemoglobin concentration in males was reported as 14.3g/100 ml of blood. This was obtained on a sample of 500 subjects. The standard deviation was reported as 1.2. In practical applications we generally use the value 2.00 instead of 1.96 as this simplifies calculation, and the error is only about 2%. From this information we can state that there is a probability of 0.95 that the universe mean is somewhere between the fiducial limits of $14.3 \pm 2(1.2/\sqrt{500})$. Note that the amount to be added to or subtracted from 14.3 is 2.4/22.4 or about 0.107, so that the fiducial limits are 14.2 to 14.4. Thus, if another investigator studying 500 subjects or even 5000 subjects reported 15.0 g as the mean, we would have to suggest either that his measuring technique was different or that the population he studied was different. We also recognize that little would be gained by studying an even larger group of subjects, as the increased amount of data would not be likely to change the estimate of the universe mean by as much as 0.1 g/100 ml or less than 1%.

We might, however, on the basis of the same data, be prepared to consider the possibility that the method used could be improved. Here the mean, 14.3 ± two standard deviations, should encompass 95% of the normal male population. This implies that 2.5% of the normal males in this group have 11.9 g/100 ml of hemoglobin or lower. (The other 2.5% are higher than 16.7.) This is a little too low to be consistent with clinical experience. Either our clinical experience or our recollection of it is invalid, or we have not used the statistical analysis correctly. Generally, while recollections of clinical experience are notoriously misleading, it is not wise to disregard them entirely. In a later chapter on the analysis of variance this problem will be explored a little more completely, as the interpretation of this kind of measurement involves both the variability of the subjects and the variability of the measurement technique that is used.

If the measurements are accepted as they stand, the implications of the questions raised should be considered in a preliminary way. In the analysis above, we have, in effect, implied that since 95% of normal males were included within the limits of $\bar{x} = 2s$, we could use 11.9 g as a lower limit of normal. Thus, if we say that any mature male with less than 11.9 g of hemoglobin/100 ml is anemic (i.e., has insufficient circulating hemoglobin), we could expect to be wrong in our evaluation of normal males about 2.5% of the time. (Note that this is not the same as saying that 2.5% of subjects diagnosed as anemic are normal.) However, in biological work it is conventional to accept a 5% risk. To find a lower limit of normalcy that incorrectly labels no more than 5% of normal mature males, we return to the tables to find a value for U that includes 90 rather than 95% of the population. Remember that we are using a table for double-tail values and that when 90% of the population is in the mid-zone, half of the remaining 10% is under the lower end of the curve, and the other half is under the upper end. In short, we are using Z instead of U as the measure of relative deviation. This value is 1.64. Then, $14.3 - (1.64)(1.2)$ gives 12.3 as the lower limit. This, incidentally, is more consistent with clinical experience.

The purpose of the preceding discussion is not to validate or justify so-called clinical experience. It is intended to illustrate in a realistic way the soundness of using the 95% level of confidence in the correct way and accepting the risk of error in 5% of the cases. In the illustration above where we first used a level leading to rejection of a true hypothesis only 2.5% of the time, we also incurred an unmeasured risk (actually much higher than 5% in this instance) in accepting the false hypothesis that subjects with 12.0 g were normal (i.e., not anemic).

In the example above we were able to set the 95% fiducial limits of the universe mean within about 1%, even though the standard deviation was about 8.5% of the mean. This was possible because there were 500 measurements in the sample. The standard error of the mean was small because of the sample size. It is not always feasible to obtain such large samples. An experimenter in biological work would be exceptionally fortunate if he did not at some time in his career encounter a study where a single measurement was more difficult to make than 500 hemoglobin determinations. Where large numbers of measurements can be made and where the variability is small, statistical analysis may only serve to confirm the obvious. Even with appreciable variability, large samples do tend to yield reliable estimates of μ and σ. In the example above we *assumed* that s was so close to σ that we were able to use it in establishing the fiducial limits. With small samples we cannot do this safely without making some allowance for the possible unreliability of s as a measure of σ. (The confidence limits for estimates of σ are considered in Chapter 11.) Essentially, to make this allowance, we multiply U by a factor, g, when setting fiducial limits of the mean. Thus, we would have

$$\mu = \bar{x} \pm (gU)s \qquad (6)$$

where the factor, g, would depend on the size of the sample. For small samples g must be larger than 1.0, but as the sample size increases, g must approach 1.0 as its limiting value. This correction factor was first solved by W. S. Gosset and published in 1908 under the pseudonym "Student." The values of (gU) for various probabilities as a function of sample size were tabulated under the heading t, and use of this corrected value is still commonly called *Student's t-test*. For samples of 30 or more, t and U are practically equal; therefore, when samples contain less than 30, we must use t in place of U. Thus:

$$\mu = \bar{x} \pm ts \tag{7}$$

Table A-3 lists t for various probability levels as a function of sample size. Note carefully that values of t are tabulated as a function of the number of *degrees of freedom* rather than as a function of the number, n, of items in the sample. The values of t are used in other applications besides establishing confidence or fiducial limits around a mean. Therefore the relationship of n to df (degrees of freedom) will be considered a little more fully below. At this point it will suffice to point out that df here is $(n-1)$. Recall that $(n-1)$ or df was used instead of n for samples of less than 30 in computing s. In any event, although the t values were designed to set fiducial limits for the means obtained with small samples, its widest application is found in statistical comparisons of two means, such as those obtained from an experimental group and a control group of data. This analysis will be presented in the next chapter.

Beginners and sometimes even more advanced students have difficulty with the concept of *degrees of freedom*. Practically every application of probability theory to statistical analysis is related to degrees of freedom rather than to the total number of values in the data under analysis. It is therefore important to obtain a clear notion or understanding of the concept.

In essence the utilization of this concept may be thought of as a device to prevent the naive mistake of using a piece of data more times in an analysis than is reasonable. Obviously, if we assumed that repeating an experiment would give identical results and therefore calculated the variance by dividing the sum of squared deviations by $2n$ instead of n, the result would be incorrect. If we insist on using n in some calculations in place of degrees of freedom, we make a comparable although more subtle mistake. What exactly then is implied by this expression?

Suppose, for example, we take the arithmetic mean of three numbers: 5, 6, and 7. The mean of this group is 6, which is also the mean of 3, 5, and 10, or of 2, 7, and 9, or of 9, 10, and −1. It would be possible to give an infinite number of three value clusters all of which have the same mean, 6. In generating these groups of numbers, we can assign any values that we wish to the first two numbers of the group, but once these two values have been assigned, we no longer have any freedom of choice in assigning the third value if the mean of the group is to remain unchanged. In other words, if we wish to duplicate the mean, we have only two degrees of freedom.

Clearly, what is true in the example above for a sample of three numbers is true for samples of any size. Thus, if we speak of the arithmetic mean of a sample consisting of 20 numbers, we can duplicate this mean with a sample of equal size; in this instance we are free to assign any value we wish to 19 of the numbers, but once this has been accomplished, there is no free choice in assigning the 20th value. In calculating the mean of a sample consisting of n numbers, we divide the sum of the numbers by n. In any further statistical manipulations where the mean as a single value represents all of the component values, we have lost one degree of freedom.

Ordinarily, the next step in the analysis of data consists of the calculation of the standard deviation. In this calculation we properly use $(n - 1)$. Now, if for a group of numbers forming a sample of size n, we were given a mean and standard deviation, we could generate a second group of size n consisting of different numbers than those in the first group. In this instance, we could freely assign values to all but two of the group and still retain the same values of the mean and standard deviation. In other words, we would now have $(n - 2)$ degrees of freedom.

We may summarize all of the above for our guidance in any analytic situation by asking ourselves how many of the numbers in the original sample could be arbitrarily changed without changing the value of the parameter being computed.

SUMMARY

The universe of discourse refers to the set of values under consideration in a statistical analysis. Generally, these sets are considered to be composed of an infinite number of values. Thus, the exact values of the parameters of the corresponding distributions are unknowable. However, as the size of a sample increases, the deviations of $\bar{x}$ from μ and of s from σ diminish. The distribution of sample means around the mean of the universe of discourse is a normal distribution. The standard deviation of this distribution is called the standard error. It is directly proportional to the population standard deviation, σ, and inversely proportional to the square root of the number of measurements comprising the sample. Thus, $SE = \sigma/\sqrt{n}$. Using the proportions encompassed by relative deviations (Table A-2), the probability of a given deviation of a sample mean from the universe mean can be calculated. Conversely, the range within which the universe mean must lie with a given degree of probability (i.e., the confidence limits) can be similarly calculated.

When the sample size is large, s may be used as a satisfactory estimate of σ in the above calculations. For samples of less than 100 measurements the above calculations are carried out with tabulated values of t rather than U to allow for the uncertainty in the use of s in place of σ. The values of t are a function of the degrees of freedom. The degrees of freedom may be defined as the number of arbitrary values that may be freely chosen in duplicating the observed parameters with a sample of the same size.

PROBLEMS

1. Using the mean and standard deviation calculated in Problem 1 of Chapter 4, estimate the 95% confidence limits of the mean. What would be the 99% confidence limits?
2. Calculate the mean and standard deviation of the 20 values given in the last column of Table 3-1. What are the 95% confidence limits of the mean?
3. A sample of 4 yields a mean of 93 and a standard deviation of 10. What is the probability that this could have been drawn from a population whose mean is 100?

6
THE NULL HYPOTHESIS AND THE
COMPARISON OF MEANS

One of the most common statistical problems in experimental biology or medicine is the comparison of the means of two samples. Thus, for example, we may need to compare the efficacy of a new therapeutic agent with an old one, or we may be asked if a certain treatment is effective at all (i.e., compared with the average results of no treatment). A zoologist may wish to determine if members of a species found on a high plateau are smaller than members of the same species found in the valley. Examples are limitless.

Statistical comparisons are not concerned only with means obtained with continuous variates. Analyses involving discontinuous variates are probably equally frequent (see Chapter 9). Not infrequently, comparisons of variability are needed. In all of these statistical problems we make use of a principle known as the *null hypothesis*.

In making comparisons of means or evaluating deviations from means, we must first decide if the observed differences are real and possibly meaningful. Faced with the inescapable variability of data, we realize that observed differences between two groups of data may be fortuitous. Thus, our statistical analysis in these situations must finally lead to a choice between two alternatives: (1) the observed deviation or difference could easily have occurred by chance, or (2) the deviation is statistically significant.

Two points must be made clear before the function of the null hypothesis in this decision is revealed. First, there are occasions when the data are not adequate to make any decision between the two alternatives. The experiment may be inconclusive, the data too variable, or the number of observations too small. Secondly, the meaning of *statistical significance* should be considered. When we state that a difference or deviation is statistically significant, all that we are implying is that the observation is not likely to be fortuitous or purely a matter of chance. The observed difference or deviation may be meaningless in the context of the experimental situation and still be statistically significant. That is, a statistically significant difference may be real, not due to chance, and still completely inconsequential.

The use of the null hypothesis is best understood by using a simple illustration. Imagine that as part of a game being played with a friend, he tosses a coin. If it comes down heads, you win a dollar; if it comes down tails, you lose a dollar. Now, the first toss comes out tails, and you accept the loss of a dollar with no great misgivings. The second toss comes out tails, and so does the third. This is annoying but not necessarily alarming. Now, suppose the fourth, fifth, sixth, and seventh tosses come out tails. At what point do you become suspicious, and at what point do you become convinced that this is not really a game of pure chance?

In the beginning of the game we started with an implied acceptance of the null hypothesis. That is, we were operating on the *hypothesis* that *nothing* other than random chance determined the outcome of each toss. After the first toss, there was no reason to reject this hypothesis, as the probability of tails occurring by chance is $1/2$. After the second and third tosses, the compounded probabilities of the events were $1/4$ and $1/8$, respectively. The probability of obtaining four tails in a row by chance is $1/16$. Still accepting the null hypothesis (i.e., the assumption of no effect other than chance), we are apt to regard the outcome thus far as unfortunate but not necessarily improbable. After the fifth successive loss, the strain on our credulity passes the breaking point. The probability of tails five times in a row is $1/32$. As mentioned, previously, statistical convention regards events with a chance probability of $1/20$ or less as improbable (i.e., outside the 95% fiducial limit). Since this fifth successive loss must be regarded as *improbable* on the basis of the null hypothesis, it is *probable* that the hypothesis is in error. We therefore reject the null hypothesis and conclude that something other than pure chance is operating.

The use of the null hypothesis is a little difficult to grasp at first, because it involves abstract statements complicated by double negatives. The source of the complexity can be seen by comparing two statements of a simple fact as follows. First, I can say quite simply that since the shoe I am trying on pinches my foot, it must be too small. However, in the roundabout phraseology of the null hypothesis, I would saw that if the shoe were not too small, it is improbable that my foot would feel pinched; therefore I must reject the hypothesis that the shoe is not too small and conclude that it is too small. Obviously, people do not talk this way.

Unfortunately, the evaluation of data is not always as easy as trying on shoes. In the dishonest coin example the "pinch" was fairly obvious. Let us see now how the roundabout reasoning of the null hypothesis is used in statistical problems.

In a way a form of the null hypothesis was applied in Chapter 5 when we used $\pm 1.96\sigma$ deviation from the mean to establish the range of normal values for hemoglobin concentration. Suppose, for example, that we have a subject whose hemoglobin concentration is only 10 g %. Using the null hypothesis, we ask, assuming that the subject does not differ from normal, what the probability is of finding such a normal individual by chance. Since 10 g % is definitely further away from the mean than 1.96σ, this probability is less than 0.05. Therefore we regard this as improbable and reject the hypothesis that the subject is normal; we conclude that the subject differs from the normal population by a statistically significant degree (i.e., he is abnormal).

The same principle is used in comparing two groups of measurements, but in this application the efficiency of the comparison may be enhanced by the design of the experiment. Thus, as shown below, if an experiment can be designed to utilize the statistical analysis of paired data instead of unpaired data, valid conclusions can be established with a smaller number of total observations.

It is generally implied that the effects obtained in an experimental group would occur in a control group if the experiment could be repeated with such a control group. In some experimental designs this type of confirmation is part of the plan, but this is not always possible. Thus, for example, if we use one method of ashing for the analysis of some animal or plant tissue and wish to see how it compares with another ashing method, we cannot reconstitute the specimens and start again with the alternate method. Essentially, we wish to avoid superimposing the variability of the material on to the variability of the method. In this example the solution is fairly simple. Each specimen, if it is not already uniform in all of its portions, is homogenized so that one portion or aliquot taken from the specimen is identical to any other aliquot. From each specimen one aliquot is used with one ashing method, a second aliquot with the other. The difference observed with each pair is then due only to the different methods of ashing.

Physical homogeneity of specimens of this kind makes the use of identical aliquots as controls fairly obvious. However, since this type of approach is so helpful in eliminating extraneous variation, we must be alert to the possibility of *pairing* observations whenever possible. Physiologic studies are often concerned with measuring the effect of some treatment, stimulus, or condition on a physiologic function. If the test treatment is nondestructive, the best experimental plan uses each subject as its own control. Measurements are made on each animal or subject twice, once under control conditions and once under experimental conditions.

The application of the null hypothesis is fairly easy when the experimental design permits pairing of the data. If we assume that there is no difference attributable to the two different conditions or treatments, the mean of the differences of the paired measurements should be zero. (Note that in this assumption we are referring to the universe mean being equal to zero.) If in the experimental data there is a mean difference other than zero, we ask what the probability is of obtaining this difference by chance. The observed difference gives an estimate of the deviation around the hypothetical mean. Table 6-1 shows in general terms a set of paired observations; x refers to the control group; x' refers to the experimental group. In each instance x is subtracted from x' to generate a population of differences, D, as shown. The mean of this sample of differences, $\overline{D}$, and its standard deviation are used to calculate t as shown in Eq. (2a).

Having obtained the parameters as indicated in the table, we now seek to determine the probability of obtaining a mean difference, $\overline{D}$, with n pairs in the sample of the universe. In a previous chapter the Student's t-table of values was used to obtain 95% confidence limits. In this application we entered the table with a previously established percent (i.e., 95%) and known degrees of freedom. Where the row (degrees of freedom) and the column (95%) intersect, we found the value of t needed. Here the question is somewhat different. We enter the table scanning the row of figures corresponding to the

Table 6-1

x	x'	$(x - x') = D$	D^2
x_1	x_1'	D_1	D_1^2
x_2	x_2'	D_2	D_2^2
x_3	x_3'	D_3	D_3^2
.	. .	.	.
.	.	.	. .
.	.	.	.
x_n	x_n'	D_n	D_n^2
		ΣD	ΣD^2

$$\overline{D} = \frac{\Sigma D}{n} \qquad s = \sqrt{\frac{n\Sigma D^2 - (\Sigma D)^2}{n(n-1)}}$$

$$t = \frac{\pm\overline{D}}{s/\sqrt{n}}$$

available degrees of freedom until we find a value closest to the value of t that we have computed from the data as shown; the heading of the column in which we find this value tells the probability of obtaining the observed difference by chance. The relationship of these two approaches is revealed in the following equations. Equation (2a) is merely an algebraic transformation of Eq. (1); but in Eq. (1) μ represents the confidence limit, while in Eq. (2a) it represents a universe mean. Since the universe mean is assumed to be zero in the present use of the null hypothesis, and the variate under consideration is $\overline{D}$, this is essentially the same as Eq. (2b).

$$\mu = \overline{x} \pm ts/\sqrt{n} \tag{1}$$

$$t = \pm \frac{(\overline{x} - \mu)}{s/\sqrt{n}} \tag{2a}$$

$$t = \pm \frac{\overline{D}}{s/\sqrt{n}} \tag{2b}$$

An illustration of this application is given in Table 6-2, which shows the results of an experiment to see if the intake of food has any effect on the rate at which the liver removes injected sulfobromph-thalein from the bloodstream. This study could have been performed

Table 6-2

Fed	Fasted	D	D^2
16.2	9.6	6.6	43.56
16.7	11.3	5.4	29.16
16.3	9.5	6.8	46.24
18.5	13.5	5.0	25.00
15.9	8.8	7.1	50.41
16.1	9.3	6.8	46.24
16.5	10.4	6.1	37.21
15.9	11.7	4.2	17.64
17.5	10.8	6.7	44.89
16.4	9.8	6.6	43.56
		61.3	385.91

$$\overline{D} = 6.13 \qquad n = 10$$

$$s = \sqrt{\frac{10(385.91) - (61.3)^2}{10(9)}} = 3.36$$

$$t = \frac{6.13}{3.36/\sqrt{10}} = 5.8$$

by measuring one group of animals in a fasting state and another comparable group after feeding. However, if we had the mean clearance measurements on two groups of animals under identical conditions, the values would not be identical. Therefore, if we observed a difference under separate conditions, the difference might be either exaggerated or obscured by this source of variability. However, in the experiment of Table 6-2 each animal has been measured twice, once while fasted and again while fed. Thus, each animal serves as its own control. Any difference due to variation between animals is thus eliminated from the final difference.

The desirability of pairing in the experimental design is fairly obvious, and elaboration of this principle underlies much of the theory of experimental design. It is not always possible to design experiments to utilize the efficiency of the pairing principle. Furthermore, the number of observations in an experimental group often is not the same as the number in the control group. Generally, in one group of n_1 observations we obtain a mean of $\overline{x}_1$ and a standard deviation of s_1. In a comparable group subjected to some experimental condition, we have n_2 observations yielding $\bar{x}_2$ and s_2 as the parameters. We then wish to determine the probability of observing the difference $(\bar{x}_1 - \bar{x}_2)$ by chance, assuming that no real difference exists between the groups.

There are several ways of approaching this problem, but not all of them are equally satisfactory. It is instructive, nonetheless, to consider them and the difficulties that they may raise. First, consider the fiducial limits of the two means; if these do not overlap, we could reasonably conclude that each mean represents a different population. Thus, the statistical significance of the observed differences would be established. However, if some degree of overlap occurred, we would not necessarily be justified in concluding that they were from the same population. In other words, we have no basis for accepting the null hypothesis. Therefore, although this approach can serve to establish the statistical significance of a difference between two observed means, it usually requires larger samples. Using the null hypothesis properly enables us to use smaller samples.

A second approach is to take the null hypothesis at face value, pool all the data, and make statistical inferences from the parameters obtained. To the statistically naive, this appears to be the most direct and simple approach, but, as will be demonstrated, this method not only provides traps for the unwary, but also has a tendency to require larger samples than the best approach.

The most efficient method proceeds as follows. Using the null hypothesis, we assume that the two samples have been drawn from the same population. Then, if we repeatedly draw pairs of samples one of size n_1, the other of size n_2 we will generate a population of differences of the means. Note that in this case *differences of means* is *not* the same as *mean of differences*. When we had paired data, we were in reality able to compute a mean difference. In the present situation we assume that *if* we repeatedly take samples of differing sizes as indicated above, these will, in the long run, generate a series of pairs with a mean of differences equal to zero. If this is the case, we now ask what the probability is of encountering as large a difference as $(\bar{x}_1 - \bar{x}_2)$ or larger with a single pair.

To answer this question, we need to know the standard deviation of this generated population of differences. We have already located μ at zero. Now we need to compute s', which is the best estimate of σ, from the samples with standard deviations s_1 and s_2. In a later chapter the generally additive nature of variances will be discussed. However, it will suffice at this point to show how pooled variance, V', is obtained from the individual variances, V_1 and V_2. Recalling that s_1 and s_2 were obtained by taking the square roots of V_1 and V_2, we may write.

$$V' = \frac{(n_1 - 1)V_1 + (n_2 - 1)V_2}{(n_1 - 1) + (n_2 - 1)}$$

In essence, the pooled variance is a weighted average of the two component variances, except that degrees of freedom are used instead of simple frequencies. By taking the square root of V', we have obtained the standard deviation, s', of our hypothetical population. Now suppose we calculate t as follows:

$$t = \frac{\bar{x}_1 - \bar{x}_2}{s'}$$

From the table of t values we find the probability of finding a difference this large between any pair of *individuals* in the pooled population equal to $(\bar{x}_1 - \bar{x}_2)$, but we need to know the probability of finding differences between means of different sample sizes. To calculate the value of t required, we use the following formula:

$$t = \frac{\bar{x}_1 - \bar{x}_2}{s'\sqrt{(1/n_1) + (1/n_2)}}$$

To illustrate this method consider the following example.

Kleiber, studying survival times in starving rats, found that some succumbed as early as 4 days, while a few survived up to 26 days or more. In an attempt to discover what conditions contributed to this variation, he considered the effect of environmental temperature. To see what effect this might have, he starved 14 rats at a constant temperature of 30°C, while 7 were starved at a temperature fluctuating closely around 20°C. The data and calculations are summarized in Table 6-3.

Although the mean survival time of the animals in the warmer environment is twice that of the control group kept near ordinary room temperature, there is a fair amount of overlap at one end of the scale. Four of the animals in the warmed group did not survive any

Table 6-3

30°C	20°C
4	4
7	8
10	8
11	10
14	10
16	12
16	<u>12</u>
21	
23	
25	
26	
26	
26	
<u>30</u>	

$$\Sigma x = 255 \qquad 64$$
$$\bar{x} = 18.2 \qquad 9.1$$
$$\bar{s} = 8.2 \qquad 2.8$$
$$V = 67.1 \qquad 7.8$$
$$n = 14 \qquad 7$$

$$V' = \frac{(13)(67.1) + (6)(7.8)}{13 + 6} = 48.4$$

$$s' = 6.96$$

$$t = \frac{18.2 - 9.1}{6.96\sqrt{(1/14) + (1/7)}} = 3.87 \qquad df = 19$$

$$P < 0.5\%$$

longer than their relatives in the cooler environment. Clearly, the doubling of the average survival time has biological significance, but the statistical significance should be evaluated. As shown in Table 6-3, the observed difference in observed mean survival times may be regarded as statistically significant, since the probability of this difference occurring by chance is less than 0.05.

SUMMARY

The null hypothesis is a method of reasoning that is used in the statistical evaluation of the differences observed between two sets of data. In applying the null hypothesis, we assume temporarily that the two sets of data arise from the same population and then determine the probability of finding the observed difference by chance. If the probability is small (i.e., less than 0.05), we reject the assumption; the only remaining possibility is the conclusion that a real difference exists. On the other hand, if the probability is higher than 0.05, we must conclude either that the null hypothesis is correct or that the data are inadequate to demonstrate a real difference. (This possibility is dealt with more fully in Chapter 15.)

In evaluating the probabilities arising from the null hypothesis, we have either paired or unpaired data. When we have paired data, the null hypothesis implies that we know the universe mean of differences to be equal to zero. The probability of obtaining the observed difference by chance is estimated from the ratio of this difference to the standard error, using U values for large samples and t values for small samples. With unpaired data the same procedure is used, only in this instance the standard error is necessarily computed from the pooled variances of the two groups.

FORMULAS

For paired data:

$$t = \frac{\bar{d}}{s/\sqrt{n}} \quad \text{df} = n - 1$$

For unpaired data with unequal samples:

$$t = \frac{(\bar{x}_1 - \bar{x}_2)\sqrt{(n_1 + n_2 - 2)n_1 n_2}}{\sqrt{[(n_1 - 1)V_1 + (n_2 - 1)V_2](n_1 + n_2)}} \quad \text{df} = n_1 + n_2 - 2$$

PROBLEMS

1. Two sets of 50 animals were raised on two different diets. The average weight of one group was 149.2 g with a standard deviation of 8.0 g; the second group showed an average weight of 152.3 g with a standard deviation of 10 g. What is the probability that the difference is because of chance variation rather than the superiority of one diet over the other? Should the hypothesis that one diet is better than the other be accepted?

2. A group of ten subjects is subject to a mild stimulus. The table below shows the blood pressure measurements, in millimeters, before and after the stimulus; the measurements are arranged in order of magnitude of the difference observed. Do these data acceptably support the hypothesis that the stimulus raises blood pressure?

Before	After
118	127
120	128
128	136
124	131
136	138
130	132
130	131
140	141
140	132
128	120

3. The experimenter who obtained the data shown in Problem 2 suggests that the subjects whose blood pressure rose less than 7 mm after stimulation (i.e., the last six listed) differed from the group whose blood pressure did rise 7 mm or more. He assumes that the relatively unresponsive subjects differed from the first four in the list in that they were mildly stimulated by the observation process and therefore could not respond further. How would one test this hypothesis? Is there a statistically significant difference between the first four subjects and the last six?

7
GRAPHS AND EQUATIONS
(REGRESSION)

"One picture is worth 1000 words." A graph is a very special kind of picture, which may be worth 10,000 more. For some of us, graphs can have a very special aesthetic quality. A well-made graph summarizes relationships, gives concrete expression to abstract equations, and may at times reveal new interpretations.

This chapter will be limited only to those graphs related to algebraic equations. The branch of mathematics which deals with this type of relationship is known as analytic geometry. This subject deals quite exhaustively with these relationships, but we shall use only those portions useful to us in statistical analysis. The need for considering a few elements of analytic geometry arises from the common sequence of processes that characterize the application of statistical analysis to biological research: (1) the accumulation of experimental or observational data, (2) the orderly tabulation of the data, (3) graphical representation of the data, and (4) summary of the data in an algebraic statement.

An algebraic equation is a concise symbolic statement summarizing quantitative relationships. Just as with a grammatically correct sentence, an equation has a subject, a verb, and a predicate. The quantities on the left side of the equation form the subject; the equals sign is the verb, and the quantities on the right-hand side of the equation form the predicate. The various indicated operations of multiplication, division, addition, and subtraction comprise the adjectives and adverbs.

Equations are generally more accurate and far more concise than tables or graphs. Thus, for example, we all know that the circumference of a circle is related to the diameter by the very simple equation, $C = \pi D$. This summarizes the relationship for all time and obviously is far more useful both practically and in theoretical derivations than a vast tabulation of diameters and related circumferences. Even if we were to use a table in some special circumstance to avoid bothersome calculations, we would frequently find it necessary to interpolate values between two entries in the table. A graph or line nomogram would then be more convenient. Thus, for example, the conversion of temperature readings on a centigrade thermometer to Fahrenheit equivalents might be awkward using a table since each step of 1 degree on the centigrade scale is equivalent to 9/5 degree on the Fahrenheit scale.

In spite of the advantages of brevity, accuracy, and simplicity, an equation by itself may be a little too abstract for easy comprehension. Theoretically, if we all had sufficient intellect, we could carry out discussions of statistical analysis by referring only to the equation of the normal curve. The symmetry and the parameters are all there. Nevertheless, a graphic picture of the distribution aids in the understanding of the relationships needed for applying the ideas of proportion and probability. To accomplish this without a graph, we would need to integrate the equation and then solve the integrated equation for the values between two levels of interest. Even with a pencil and paper this would be a difficult task. Using the bell shaped curve, we can see quite readily the nature of a problem and then are enabled to select the appropriate tables.

One of the most common procedures in scientific investigation is the quantitative study of the influence of one variable on another. As has been indicated, one of the most illuminating ways of presenting results is with a graph. This task is accomplished by following certain established conventions. The magnitude of one variable is indicated on a horizontal scale, *the abscissa*, while the other variable is indicated on a vertical scale, *the ordinate*. These terms for the vertical and horizontal scales are widely used, but occasionally students have some difficulty in remembering the difference between one and the other. Therefore, it may be helpful to point out that the word "ordinate" has the same root for its derivation as the word "ordinary," which literally means straight (i.e., not deviating to one side or the other). The word "abscissa" is related in its origin to the word "scissor." The horizontal abscissa quite literally cuts across the vertical ordinate. When the lines are considered together, they are referred to as *the coordinates*.

Since each coordinate refers to one of the pair of variables under consideration, it is customary to use the abscissa for the independent variable and the ordinate for the dependent variable. Generally, one variable either changes or grows regardless of what we do or else is a variable over which we have relatively complete direct control. These are *independent variables*. The quantities which change as a result of the change in the independent variable are then the *dependent variables*. Many of the phenomena which we study change with time (e.g., height, income, population, etc.). In these studies time is obviously the independent variable; its effect would be the dependent variable. If we study the effect of varying doses of a drug, the dose would be the independent variable; the effect would be the dependent variable. When algebraic relationships are being considered in generalized terms, we use the letters, x and y for the independent and dependent variables, respectively. In algebraic equations it is also customary to use upper case or capital letters for the constants or parameters and to reserve the small or lower case letters for the variables.

The process of transforming data to a graph is not difficult or complicated. Although there are exceptions, generally the distinction between independent and dependent variables is fairly obvious. The only part of the task that requires any judgment is usually the selection of appropriate scales for the coordinates. This judgment is generally based on a reasonable compromise between convenience and clarity, with particular emphasis on the latter.

Once all of the individual points (each of which corresponds to the individual measurements or observations) have been plotted, the nature and direction of the trend of the data should be relatively clear. It would be unusual for the points to clearly delineate a particular line or shape. That is, a line joining all of the points in sequence is likely to form a ragged or saw-tooth pattern. We recognize, however, that most processes or phenomena in nature show smooth rather than abrupt transitions. Therefore, a smooth line, which will touch or come as close as possible to all of the points in the array, is desired. This requirement may be satisfied best by a curve or a straight line. Most of the available statistical techniques, which are commonly employed, are designed to locate on the graph the straight line that fits the data better than any other straight line.

Where the data obviously are part of a curve, the usual strategy is to modify the mathematical expression of the data in such a way as to convert it to a straight line. Before these statistical methods can be discussed intelligibly, the relationships between straight lines and the class of algebraic expressions called *linear equations* must be considered.

The simplest linear equation takes the form:

$$y = Ax \tag{1}$$

In this equation there is a simple, direct proportionality between the values of the dependent variable, y, and the independent variable, x. When x is equal to zero, y is also zero. A simple, nonstatistical example should suffice to make this type of relationship clear. Suppose we wish to set up an equation expressing the relationship of body weight in pounds to the same weight in kilograms. In this case we will arbitrarily choose to designate weight in kilograms as x, the independent variable. The dependent variable, y, will then equal the weight in pounds. Thus:

$$y = 2.2x \tag{2}$$

Starting with zero, we could list a series of weights in pounds and then calculate the corresponding weights in kilograms. The resulting table would look like this:

x	y
0	0
1	2.2
2	4.4
3	6.6
4	8.8

Note that in Eq. (2) the value, 2.2, has been substituted for A in Eq. (1). Figure 7-1A shows the graph that would correspond to the equation covering the tabulated values. One of the great advantages in dealing with linear equations and straight lines is illustrated in Figure 7-1B. This graph was constructed with far less labor, using a

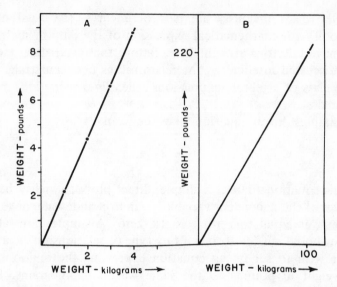

Figure 7-1. (A) Weight in pounds as a function of weight in kilograms. (B) Line established by a single point and the origin.

single point and zero to establish the location of the line. Using rules to be summarized below, we recognize the equation as being linear. We also can recognize that the line for the equation must go through the origin (i.e., the point of intersection of the coordinates where both scales are equal to zero). Since only two points are needed to establish the location of a straight line, and since we already know the origin to be one of the points, we need only calculate one more point. In this way we obtain in Figure 7-1B the line covering the range of weights in kilograms from 0 to 100 without the bother of constructing a table.

Although Eq. (1) is the simplest form of a linear equation, it is best thought of as a special case of the general equation for a straight line:

$$y = Ax + B \tag{3}$$

In this equation B stands for either a parameter or constant. Like the parameter A, it may be either positive or negative. As a simple example of how an equation of this type works, assume that we have a scale calibrated in kilograms, but that the attachment of the indicator is defective and points to −0.27 kg when the weighing platform is empty; if we put an object weighing 0.27 kg on the platform, the indicator will point to zero. We now wish to express in an equation the relationship of the weight of an object in pounds to its *apparent* weight in kilograms on this defective scale. We would write:

$$y = 2.2(x + 0.27) = 2.2x + 0.6 \tag{4}$$

The graph of this equation is shown in Figure 7-2A. Equation 2 is also shown as the dotted line on the same graph. Notice that these two equations form a pair of parallel lines. This is a consequence of their sharing the same coefficient, 2.2, in front of the independent variable. This value determines the steepness of the line, and therefore, the value, A, in Eq. (1) is referred to as the *slope* of the equation.

To carry this illustration one step further, suppose we have a scale similar to the previous one but without a defective pointer and with a chair bolted to the scale platform so that subjects may be weighed in the sitting position. If the chair and its attachments weigh 1.4 kg, and we wish to express the correct weight of the subject in pounds in an equation as a function of the indicated weight, we would write:

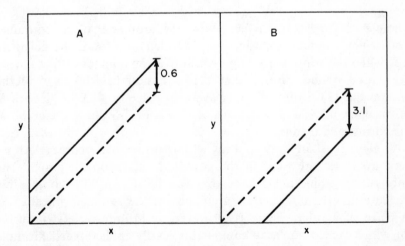

Figure 7-2. (A) Graph showing location of lines for $y = 2.2x + 0.6$ and for $y = 2.2x$. (B) Lines for $y = 2.2x - 3.1$ and $y = 2.2x$.

$$y = 2.2(x - 1.4) = 2.2x - 3.1 \tag{5}$$

Figure 7-2B shows this line along with the line for Eq. (2) again shown as a dotted line. Note that the equations again form parallel lines, but this time, unlike the line for Eq. (4) which lies above that of Eq. (2), this line lies below the dotted line. The distances which these lines lie above or below the dotted lines are in each instance determined by the value of the parameter, B, of Eq. (3). These examples illustrate how this parameter may have either positive or negative values. Thus, Eq. (1) is considered to be a special case of Eq. (3) in which $B = 0$.

The parameter, A, may take on negative as well as positive values, but obviously cannot take the value zero except to imply that y is completely independent of x. A graph of data obtained under the circumstances of this special case would form either a horizontal line parallel to the abscissa or a random scattering of points. When A is positive, a graph of the corresponding line slopes upward to the right. This is called a positive slope. If A is negative the line slopes downward to the right and is called a negative slope. A simple illustration of this case is afforded by expressing the height above the ground of a container suspended on a spring as a function of the weight in the container. Then,

$$h = K - Gw \tag{6}$$

where h is the height in feet above the ground that the container rests when a weight in pounds, w, is added to it. K is the height in feet when the container is empty, and G is the number of feet moved per added pound. Notice that this equation has the *form* of the general Eq. (3). In Eq. (6) we have h in place of y, K in place of B, $-G$ in place of A, and w in place of x. Figure 7-3 shows the graph of the corresponding line.

Straight lines are easy to draw with considerable accuracy. All we need do is set the ruler in the correct location on the graph. The only other geometrical shape that can be drawn with comparable perfection is the circle. Here a compass serves as a mechanical guide in place of a ruler. There is, however, an infinite variety of curves for which we do not have comparable, easily applied mechanical devices. Often the mathematical problems encountered in dealing with their corresponding equations are equally difficult. A few relatively

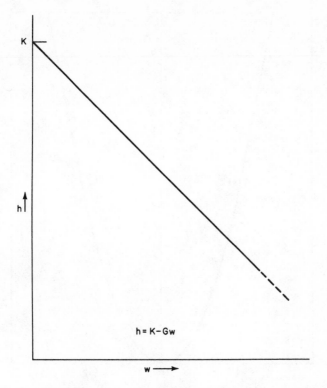

Figure 7-3. Graphic relationship of parameters of a line with a negative slope.

simple examples will demonstrate these problems as well as the technique employed to solve them.

Suppose we have an equation of the form:

$$y = Ax^2 + B \qquad (7)$$

In the table below a few pairs of values are tabulated so that we may plot the points and see what shape is obtained. Remember that negative values give positive numbers when squared. The negative values are plotted by extending the abscissal scale to the left of the ordinate. In the same manner, when negative values are encountered for y, we plot them by extending the scale of the ordinate below the level of the abscissa. For simplicity let $B = 0$ in this example.

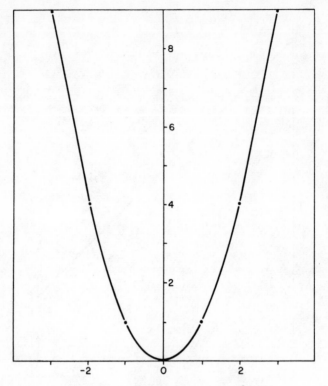

Figure 7-4. Parabola for $y = Ax^2$.

x	y
1	1
−1	1
2	4
−2	4
3	9
−3	9

The curve formed by these points is shown in Figure 7-4 and is known as a parabola.

Another curve form known as a hyperbola arises from an equation of the form:

$$y = \frac{A}{x} + B \tag{8}$$

Using the values 4 and 6 for the parameters, A and B, in this equation, the student can easily tabulate a few representative values and plot them. The graph of this equation is shown in Figure 7-5. In this graph pairs of values were used where both x and y took on negative as well as positive values.

The difficulty in drawing curves for Eqs. (7) and (8) can be obviated by transforming the independent variables in these equations. In Eq. (7), for example, we could let $z = x^2$, and substitute in Eq. (7) to obtain the same form as Eq. (3) as follows:

$$y = Az + B \tag{9}$$

The same maneuver can be used with Eq. (8) by letting $z = 1/x$. The utility of these maneuvers as an aid to determining the parameters of equations derived from experimental data will become clear as we continue.

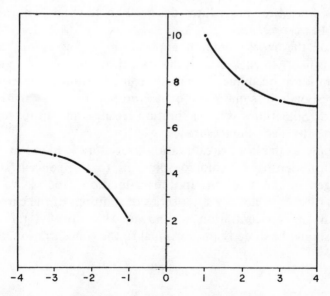

Figure 7-5. Hyperbola $y = A/x + B$.

Thus far, simple, nonstatistical examples of linear equations have been used. When experimental data showing the relationship of one variable to another are to be analyzed, there are two general classes of problems. In one instance, we know from the nature of the problem the character of the equation which the line will fit. In the second instance, we know only that a relationship exists, but the nature of the equation must be defined from the shape of the curve. Whenever the variables can be written on opposite sides of the equation as numerators without exponents, we have what is known as a *linear equation of the first degree.* In the language of statistics the analysis of this type of data is referred to as *regression analysis.* The line is called a *regression line* and the parameters are called *regression coefficients.* The term "regression" is an unfortunate one, but it continues in usage as a vestigial reminder of some of the earliest applications of statistical analysis to biological data. It was first used by Galton toward the end of the nineteenth century to express a tendency of offspring to return or regress to a mean value of the attributes of the parents. At any rate, as the expression is used today, it applies to the problem of fitting a straight line to data where one variable is well known to be dependent on the other variable. The term "regression" is used in contrast to the situation where we only suspect that there may be a relationship between two variables. Statistical procedures designed to test this hypothesis are referred to as *correlation.*

One of the most common applications of regression that the student is likely to encounter is the statistical evaluation of constants in calibration procedures. Generally, some instrument or measuring device responds in some more or less proportional way to the item measured. Sometimes the calibration relates an entire series of procedures to the item measured.

A simple example of a calibration procedure is found in relating the optical density of colored solutions to the concentration of chromogen in the solution. In the calibration process, we obtain readings of optical density on a series of solutions of the chromogen in which the concentration is known. Theoretically, the optical density should be directly proportional to the concentration, so that:

$$y = Kx \tag{10}$$

Because of variations which creep into the process of measuring optical density, the data are apt to look like the scatter shown in

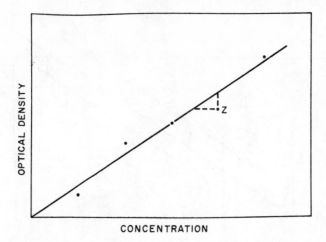

CONCENTRATION

Figure 7-6. Measurements of optical density as a function of chomogen concentration.

Figure 7-6. Presumably, the variations which produce this scatter are random, small, and produce deviations on both sides of the line with about the same frequency and order of magnitude. This statement implies that for each value of x, which we might choose, there exists at that level a population whose mean lies on the line described by the equation. The standard deviation of each of these populations is the same at each level. Figure 7-7 shows two such populations with their means lying on the line of the equation and the standard deviations producing an equal spread for each of these populations along the ordinate. In this figure which shows the spread of points relating optical density to concentration, line 1 is the line that best fits the data. Lines 2 and 3, which lie parallel to line 1, represent the lines connecting the standard deviations of all the populations of points which might lie along line 1.

The process of fitting a straight line to a scatter of data consists of finding a line for which the observed data have the highest probability. This statement is clarified by Figure 7-8. In this figure we see a scatter of points around line 1. If line 1 represents the true relationship between x and y, there is a certain probability that the observed points would be obtained by chance. At the moment we do not know what the probability is, but we shall let P_1 represent this probability. Now consider the alternative hypothesis that line 2 represents the true relationship between x and y. On this basis there

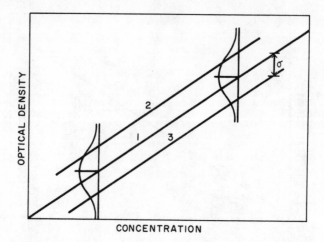

Figure 7-7. Populations of optical density measurements.

would be some other probability, P_2, that the observed points would be obtained by chance. We can easily see that P_2 must be very much less than P_1. We would be justified in concluding that if line 2 were the "true" line, the points obtained would have such a small probability of occurring by chance that we would be justified in rejecting this hypothesis. (Note the similarity between the reasoning here and that used with the null hypothesis.)

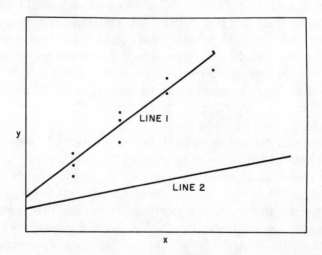

Figure 7-8. Probable (1) and improbable (2) lines for experimental data.

In this illustration the alternate line is so far away from the points that we have no difficulty in rejecting it in favor of the first line. If line 2 were fairly close to line 1, we would need a more quantitative basis for such a decision. This is accomplished by placing line 1 in a position with respect to the points so that any other line would yield a lower probability of obtaining the points by chance.

The underlying principles of this method are illustrated in Figure 7.9. In this graph only a single set of y values are shown, all of which have the same value for x. As indicated previously, these points are assumed to have a normal distribution about a mean level, indicated as μ in the figure. The distribution is shown projected on to the ordinate. If the standard deviation of this distribution is s, the probability of each of the points occurring by chance can be determined from the relative deviation. Thus, we could find $U_1 = (\mu - y_1)/s$, $U_2 = (\mu - y_2)/s$, and so on up to U_n if there are a total of n points. From the appropriate table we could then find the corresponding probabilities, $p_1, p_2, \cdots, p_n$. The compound probability associated

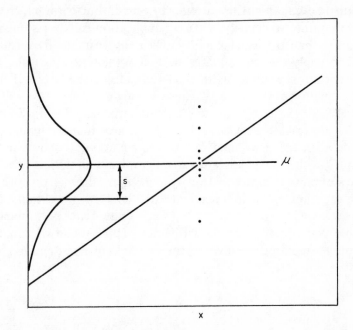

Figure 7-9. Projected frequency distribution of a population of measurements at a single level of the independent variable.

with the particular combination of points obtained would then be P_1 = $(p_1)(p_2)(p_3) \cdots (p_n)$. Now, it is obvious that P_1 will be at its maximum when all of the component p values are maximum. This will occur when the mean is so located that the variance calculated on this basis is at a minimum. Note that in this instance we require that each of the U values calculated will be maximum, as we are considering the proportion of the area under the normal curve included rather than excluded by the limits set by the relative deviations.

From the preceding considerations, it follows that if the variance is to be at a minimum, the sum of the squares of the deviations of the points from the line must be at a minimum. The process of accomplishing this is referred to as fitting the line by the *method of least squares*. Deriving the formula for fitting lines by the method of least squares is a standard exercise in calculus courses demonstrating the application of partial differentials. This demonstration will not be repeated at this point, but the general formula will be given at the end of the chapter. Formulas, however, are easily forgotten and difficult to find when they are most needed. Furthermore, they are easy to confuse, especially when an application calls for some notation other than the usual x and y. Over the years we have found it useful to apply the least squares method in the following way. First, the equation is written in its final form, using the letters A and B for the unknown parameters. Thus, we might have $y = Ax + B$, or c = $Ad + B$, or the letters for any pair of variables. The only precaution we need observe at this stage is to be sure that the independent variable is on the right-hand side of the equation, as shown in these examples, while the dependent variable is placed on the left-hand side. We now obtain a second equation by multiplying through on both sides of the equation by the independent variable. Thus, we obtain, for example, $xy = Ax^2 + Bx$ or $dc = Ac^2 + Bc$. We now write two summations of these two equations as shown below. This yields two equations, the number required to solve for the two unknowns, A and B.

$$\Sigma y = A\Sigma x + nB \tag{11}$$

$$\Sigma xy = A\Sigma x^2 + B\Sigma x \tag{12}$$

These equations are then combined or used in whatever way is most convenient at the time for solving these simultaneous equations. Once we have determined the values of the parameters A and B in this way, and obtained the equation in its final form, we may wish to evaluate the reliability of the calibration.

Before the details of this evaluation are discussed, note that in this example we used the data of a calibration line that we knew ahead of time would be best described by an equation of the form $y = Ax$. This is a special form of the general equation $y = Ax + B$, which is easily fitted to a least squares equation by the special equation $A = \Sigma xy/\Sigma x^2$. For all cases the following procedure and considerations are identical.

In Figure 7-6 the point marked z shows the importance of distinguishing between dependent and independent variables in the way in which the equation finally will be used. In setting up the initial calibration relationships, we might think it obvious that the known, predetermined concentrations would constitute the independent variables and the resulting observed optical densities would constitute the dependent variables. However, in the final application of the formula summarizing the relationship, we would be using optical density to calculate the concentration. In this application, the optical density would become the independent variable. If we treat the equation in the original way, we can see by referring to Figure 7-6 that we would want to make the sum of the deviations from the line in the lateral direction a minimum rather than the squares of the deviations in the vertical. This is accomplished by letting y stand for concentration and x for optical density.

In evaluating the reliability of the calibration, we are in essence asking what the limit of the error may be so that is is exceeded less than approximately 5% of the time. In other words, at a given optical density (i.e., at a given value for x) what is the standard deviation for the population of concentrations (i.e., values of y) that will be encountered? As will be recalled, we have already assumed that the standard deviation is the same at all levels of abscissal values. Therefore the deviation encountered at one level is comparable to a deviation at any other level, and we may use the squares of the deviations at all levels. We can find this sum of squared deviations by obtaining a calculated value, y', for each value of x, substracting this from the experimentally observed value, y, squaring each of these deviations, and then adding them up to calculate the variance. This sum divided by $(n - 2)$ is the variance about the line. The square root of this variance is the standard deviation around the line known as the standard error of the estimate (SEE). Using the approximate value for U in Table A-2, we can now state that fewer than 5% of the estimated concentrations will deviate from the calculated value by more than $1.96\ s$. Of course, the same limitations apply here as in other applications when the sample size is less than 30. It is then better to use the corresponding t value instead of U. Note that in calculating the equation of the line, two degrees of freedom are lost.

SUMMARY

Linear equations have the form $y = Ax + B$. Whenever it is necessary to fit data to a more complex equation yielding a curve, the statistical problem can be simplified by converting the equation to a linear form by appropriate substitutions. A linear equation is characterized by having the independent and dependent variables as numerators on opposite side of the equation without exponents.

The method of least squares is used to fit a line to data in which the dependent variable is a linear function of the independent variable. This method minimizes the variance of the data about the line and thus represents the most probable locus of the line. The variance of the points around the line yields a standard deviation referred to in this application as the standard error of the estimate. It may be used to evaluate the reliability of predictions based on the equation of the line.

FORMULAS

Equations for determining the parameters by the least squares method for a line described by the equation: $y = Ax + B$:

$$A = \frac{n\Sigma xy - \Sigma x\Sigma y}{n\Sigma x^2 - (\Sigma x)^2} \qquad B = \frac{\Sigma x^2\Sigma y - \Sigma x\Sigma xy}{n\Sigma x^2 - (\Sigma x)^2}$$

Standard error of the estimate:

$$SEE = \sqrt{\frac{\Sigma(\text{dev. from line})^2}{(n - 2)}}$$

where n = number of points on the graph.

PROBLEMS

1. From the following sets of data draw suitable graphs and find the equations of the lines:

a. y	x	b. y	x	c. y	x
3	1	10	2	41	9
9	3	13	3	35	15
21	7	19	5	29	21
24	8	31	9	23	27

2. When a calibration curve for the estimation of prothrombin concentration is to be formed, plasma from a normal subject is regarded arbitrarily as having a concentration of 100%. Serial dilutions of this sample with saline are prepared, and the time required for clotting after adding a mixture of calcium chloride and thromboplastin is measured. The following table illustrates the kind of data obtained:

Nominal prothrombin concentration, y	Time, seconds, required for appearance of clot, x
100	14
50	18
25	22

   a. Plot y as a function of x and draw a smooth freehand curve through the points. Use this graph to estimate the prothrombin concentration of an unknown specimen that requires 20 seconds to clot.
   b. "Rectify" the data by plotting $1/y$ as a function of x. This will fit a linear equation of the form $1/y = Ax - B$. From the graph and data find the values of A and B. Using this equation, calculate the concentration of the specimen requiring 20 seconds to clot. Compare this result with the graphic estimation in part a.

3. During the calibration of a photocolorimeter, known concentrations at 5, 10, 15, and 20 mg % were read with corresponding optical densities of 0.22, 0.31, 0.38, and 0.47.
   a. Find the equation for calculating the concentration from the optical density.
   b. Calculate the SEE.
   c. On the basis of these data estimate the 95% confidence limits for measurements with this instrument.

8
CORRELATION

In the previous chapter the problem of fitting data to a straight line was considered. In the examples considered we knew beforehand that one function was dependent on the other. Each of the examples concerned fairly obvious relationships. Thus, for example, it was obvious that optical density would be related to concentration. The existence of linearity in the relationship may not always be so certain, but in most problems of this kind the existence of a definite relationship of some kind between the dependent and independent variables is quite clear.

There are many occasions in which a relationship is suspected but not known for certain. Thus, for example, we might for one reason or another wonder if weekly income is related to height. If it were, we could plot the data showing income as a function of height. Then, using the methods of the previous chapter, we could derive a formula allowing a prediction of a man's income from his height, or, from the added use of the standard error of the estimate, the limits within which his income must fall. Now even without accumulating factual data, most of us are likely to agree that the predictive value of this formula would be poor in an industrialized society with a large middle class containing many professional as well as skilled workers. This would be reflected in our analysis of the data by two things: (1) the slope of the line would be fairly shallow (assuming there was any slope at all in the data), but more significantly, (2) there would be a wide scatter about the line (i.e., the SEE would be large). The data probably would look like the graph shown in Figure 8-1A.

Conversely, if we examined some data relating years in school to weekly income, the configuration would be more like that shown in

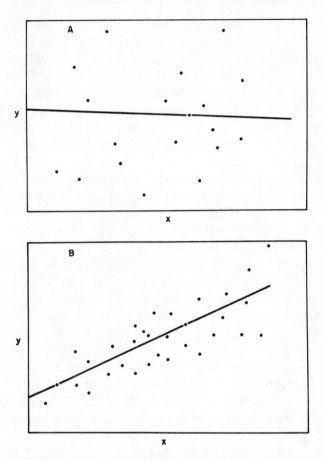

Figure 8-1. (A) Scatter of points showing no correlation. (B) Scatter of points showing definite correlation.

Figure 8-1B, with a more definite slope and much less scatter. The problem we face in the statistical analysis of this kind of data is the determination of the presence or absence of *correlation* between the two variables. In other words, we need to determine whether or not the correlation shown is real and significant.

The paragraphs above imply that a shallow slope and large standard error of the estimate would be against the hypothesis that a significant correlation existed, while a steep slope and narrow scatter of data about the line would favor the hypothesis. These, however, are at best only semiquantitative statements. Furthermore, these parameters in their absolute terms would not give us a basis for comparison nor would we be able to utilize some form of the null hypothesis. Both of the values (slope and standard error of estimate) obtained up to this point will depend on the units used. For example, if pennies are used as the units for expressing weekly income rather than dollars, the standard error of the estimate in both of the examples above will appear to be very large. In short, we need a dimensionless or purely abstract number.

This is accomplished by using the *coefficient of correlation* as first proposed by A. Bravais and K. Pearson. This yields an abstract number which gives a quantitative evaluation of how well the two values are correlated and may be used to estimate the statistical significance of the apparent correlation.

Theoretically, a satisfactory dimensionless *index* of correlation could be obtained by dividing the standard error of the estimate, SEE, by the standard deviation of the dependent variable s'. The value for SEE would be chosen as the numerator rather than the denominator for such an index, as this would limit the possible range of values from a minimum of 0, when there was perfect correlation, to a maximum of 1.0, when there was no correlation at all. If SEE were used as the denominator, the values would approach infinity as the correlation approached perfection. However, although the ratio SEE/s' could be used, this index ratio would be cumbersome in practice.

The correlation coefficient that is generally used in statistical analysis is a practical simplification of the above ratio. If r is the coefficient of correlation, the relationship to the index ratio described above is as follows:

$$r^2 = 1 - (\text{SEE})^2 / s'^2$$

This can be simplified to the formula:

$$r = \frac{n\Sigma xy - \Sigma x \Sigma y}{n^2 x's''}$$

where s' is the standard deviation of y, the dependent variable, and s'' is the standard deviation of x, the independent variable.

When this coefficient of correlation is used, a value of zero indicates no correlation at all, while the limiting values, +1 and −1 indicate either perfect positive correlation or perfect negative correlation. The probability of obtaining the calculated coefficient by chance is given in Table A-4. Note that this table is entered for various levels of probability through the column giving the degrees of freedom; after the coefficient of correlation has been calculated in the usual way, the degrees of freedom are equal to $n - 1$, where n is the number of points appearing on the graph.

It is important to emphasize that the demonstration of a statistically significant coefficient of correlation does not necessarily establish a valid relationship between two sets of values. Two sets of values could quite possibly vary in the same direction at the same time without there being any direct relationship between them. One must be particularly careful in deducing a cause and effect relationship on the basis of correlation. For example, it would probably be easy to obtain what appears to be a significant coefficient of correlation between the number of television sets sold in the period between 1955 and 1960 and the admissions to mental institutions. While this might tempt one to conclude that one phenomenon could be the cause of the other, it is far more likely that both are related to the general level of the economy of the country; with sufficient technical development and prosperity to allow a widespread increase in the number of television sets, there is usually a sufficient increase in the income from taxes to build more mental hospitals.

The most common error of this kind that we encounter in the application of the coefficient of correlation is the comparison of two sets of measurements that are both closely related to the passage of time but do not bear any direct causal interrelationship.

While it is important not to conclude mistakenly that two sets of values are related, it is equally important not to reject a valid relationship. This can easily occur if the coefficient of correlation is more or less blindly applied to the available data. A valid relationship may then be obscured if the relationship between the variables is not fairly close to being linear. All of the derivational steps in formulating the coefficient are based on the assumption of a simple linear relationship and thus may not be properly applicable. There are also statistical problems in which ordinal numbers rather than cardinal numbers appear. *Cardinal numbers* are the numbers generally used to indicate magnitude in terms of definite units; *ordinal numbers* indicate relative position in a sequence. Problems of this type are analyzed by the method of *rank correlation*.

In applying the method of rank correlation, we have pairs of associated measurements, $x_1y_1, x_2y_2, \cdots x_ny_n$, and must determine whether or not the y measurements are correlated with the x measurements. This method is accomplished by first arranging the data in a column of x values in order of their magnitude. Each of these values is assigned a rank value equal to its place in the sequence. Thus, the first in the column is assigned the value 1, the second is assigned the value 2, and so on. The y values are then also ranked and assigned rank numbers, independently. Next, the x rank number is subtracted from the y rank number yielding a difference, d. We then calculate the rank correlation coefficient, r', from the following formula:

$$r' = 1 - \frac{6(\Sigma d^2)}{n(n^2 - 1)}$$

The r' values for the rank coefficient of correlation are similar to the coefficient of correlation described previously in that the values range from zero with no correlation to +1 or −1 with perfect positive or negative correlation. Tables of significant values are given as Table A-5.

Both Tables A-4 and A-5 are used in testing the null hypothesis applied to correlation as follows. When we have obtained the appropriate coefficient of correlation, we set up the null hypothesis by assuming that the two sets of data are independent (i.e., they are not related). We then ask what the probability is of obtaining the observed coefficient by chance if the data are independent. If the probability is very low (i.e., less than 5%), we reject the hypothesis of independence and conclude that there is a statistically significant correlation.

The application of these methods of correlation and regression are generally applied in a sequence just the reverse of the sequence in which they have been presented. The minor but interesting niceties involved in their use can be best revealed by a suitable example. For this purpose, reconsider the data presented on sulfobromphthalein clearance in fed and fasted rabbits. Inspecting the data in Tables 8-1A, B, C, and D, we may think that the fall in clearance is related to the high level after feeding. However, just to be certain that we

Table 8-1A

No.	x· Fed, ml/min/kg	y Fasted, ml/min/kg	z Difference
1	13.7	8.5	5.2
2	14.2	9.8	4.4
3	14.2	12.1	2.1
4	21.2	10.1	11.1
5	16.9	7.3	9.6
6	14.6	12.8	1.8
7	16.7	9.1	7.6
8	16.4	10.1	6.3
9	21.3	9.5	11.8

Table 8-1B

Ranked	No.	Ranked y	No.	Ranked z	No.
13.7	1	1.8	1	7.3	1
14.2	2.5	2.1	2	8.5	2
14.2	2.5	4.4	3	9.1	3
14.6	4	5.2	4	9.5	4
16.4	5	6.3	5	9.8	5
16.7	6	7.6	6	10.1	6.5
16.9	7	9.6	7	10.1	6.5
21.2	8	11.1	8	12.1	8
21.3	9	11.8	9	12.8	9

do not overlook the possible unsuspected existence of physiologically significant data, we shall at the same time investigate the possible relationship of the magnitude of change to the fasting level of clearance. We have, in order to accomplish this comparison, labeled the columns x, z, and y. In Table 8-1B the data of each of the columns are rearranged in order of their size, that is, they are ranked. They are now assigned their ordinal numbers. Note that when there are two numbers in the ranked series of cardinal numbers that are the same, they are each assigned the same average number. Thus, instead of the second and third numbers in the ranked x series being arbitrarily

Table 8-1C

No.	x-rank	y-rank	d	d²
1	1	4	−3	9
2	2.5	2	0.5	0.25
3	2.5	3	−0.5	0.25
4	8	8	0	0
5	7	7	0	0
6	4	1	3	9
7	6	6	0	0
8	5	5	0	0
9	9	9	0	0

$$\Sigma d^2 = 18.5$$

$$r' = 1 - \frac{6(\Sigma d^2)}{n(n^2 - 1)} = 1 - \frac{6(18.5)}{9(80)} = 0.846$$

assigned the values 2 and 3, they are both assigned the value 2.5. Similarly, if the fourth, fifth, and sixth numbers had been the same, their ordinal values would all be 5.

Table 8-1C gives the alignment of the x rank and y rank numbers, the column of differences, d, and the column of squared differences, d^2. The summation and succeeding calculation of r' is shown at the bottom, and in this instance the value is 0.846. Using the null hypothesis, we assume that the values are independent of each other. Based on this assumption, we ask what the probability is of this r' value occurring by chance. We look up the appropriate value in Table A-5 and find that with $n = 9$, r' could be as large as 0.600 5% of the time and as large as 0.783 only 1% of the time. Since a value as large as 0.846 could occur by chance less than 1% of the time, we must reject the hypothesis that these values of x and y are independent of each other; we conclude that there is a statistically significant correlation.

Table 8-1D shows the same analysis for z ranked and y ranked values. In this case r' has the value −0.521, suggesting the possibility of an inverse or negative relationship. Nevertheless, we proceed as before by assuming that the values have no relationship and are independent and ask what the probability is of fortuitously obtaining a value of 0.521. The table indicates that this is greater than 5%.

Table 8-1D

z-rank	y-rank	d	d²
1	7	6	36
2	4	2	4
3	6	3	9
4	9	5	25
5	3	2	4
6.5	8	1.5	2.25
6.5	5	1.5	2.25
8	2	6	36
9	1	8	64

$$\Sigma d^2 = 182.50$$

$$r' = 1 - \frac{1095}{720} = -0.521$$

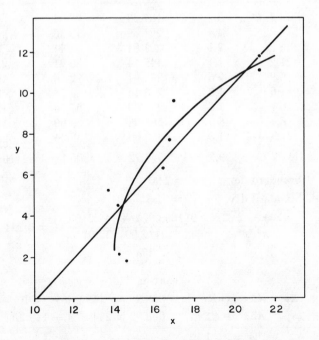

Figure 8-2. Trial comparison of straight line and free-hand curve to represent data.

Therefore, we continue to accept the hypothesis that the values vary independently of each other.

Returning now to the set of values where there was a significant rank correlation, we plot the values of y as a function of x as shown in Figure 8-2. We try fitting both a freehand curve and a straight line to the data to see which is more likely to represent the data adequately. A certain amount of "art" enters into the freehand delineation of a curve, but with very little practice a surprisingly good curve can be drawn. Even when these appear fairly crude, they are adequate enough to reveal the form of the equation that should be tried in transforming the data to a linear form. In this instance we can see by inspection that a straight line fits reasonably well. The straight line in this instance is a "fit by eye" and is done by laying a ruler over the

Table 8-2

			PART I		
No.	x	y	x^2	y^2	xy
1	13.7	5.2	187.69	27.04	71.24
2	14.2	4.4	201.64	19.36	62.48
3	14.2	2.1	201.64	4.41	29.82
4	21.2	11.1	449.44	123.21	235.32
5	16.9	9.6	285.61	92.16	162.64
6	14.6	1.8	213.16	3.24	26.28
7	16.7	7.6	278.89	57.76	126.92
8	16.4	6.3	268.96	39.69	103.32
9	21.3	11.8	453.69	139.24	251.34
	149.2	59.9	2540.72	506.11	1068.92

Standard dev. $x = s' = 2.90$

Standard dev. $y = s'' = 3.66$

$$r = \frac{n\Sigma xy - \Sigma x\Sigma y}{n^2 s' s''} = \frac{9(1068.92) - 149.2(59.9)}{81(2.90)(3.66)} = 0.794$$

$$p < 2\%$$

PART II

$$\Sigma y = A\Sigma x + nB \qquad 59.9 = 149.2A + 9B$$

$$\Sigma xy = A\Sigma x^2 + B\Sigma x \qquad 1068.92 = 2540.7A + 149.2B$$

$$A = 1.13$$

$$B = -12.05 \qquad y = 1.13x - 12.05$$

data so that an approximately equal number of points fall on either side of the line.

Having decided on a linear form for the data, we may proceed to test the data by the more appropriate correlation coefficient. The steps in this procedure are shown in part I of Table 8-2. This added step of checking by the Bravais-Pearson correlation coefficient may in this instance appear to be a waste of time, but as shown in part II, the same values are needed to calculate the parameters of the line calculated by the method of least squares. These values are also needed in the final step of calculating the standard error of the estimate. Furthermore, a significant coefficient of correlation confirms the decision to use a straight line rather than a curve.

SUMMARY

The method of least squares described in Chapter 7 is properly applied only after it has been shown that (1) the two variables are correlated, and (2) the relationship between them is linear. The statistical demonstration that two variables change together does not necessarily prove that they are directly related. Unless a linear relationship is more or less obvious from the nature of the problem, correlation should first be demonstrated by the rank method which is independent of the assumption of linearity. Each set of variables is ranked independently; the difference in rank number between the members of each pair of associated variables is squared; these squared differences are added up to obtain Σd^2 and used in the formula below. The coefficient of correlation obtained with the formula below tests for linear correlation. With both the rank coefficient and the coefficient of correlation a final value of 0 shows complete independence of the variables, while a value of 1.0 demonstrates perfect correlation. The probabilities of obtaining the observed coefficients by chance are tabulated as a function of the number of pairs of values.

FORMULAS

Rank correlation coefficient:

$$r' = 1 - \frac{6(\Sigma d^2)}{n(n^2 - 1)}$$

Coefficient of correlation:

$$r = \frac{n\Sigma xy - \Sigma x \Sigma y}{n^2 s' s''}$$

where s' = standard deviation of x, and s'' = standard deviation of y. In both of the above formulas n is the number of points on the graph.

PROBLEMS

1. Consider the blood pressure data in Problem 2 at the end of Chapter 6. If we consider the change in blood pressure to be related to the initial blood pressure, what is the rank coefficient of correlation?
2. Calculate the Bravais-Pearson coefficient of correlation on the same data, and fit an equation and line by the method of least squares.

9
ENUMERATION STATISTICS

The preceding chapters were concerned almost exclusively with the statistical analysis of continuous variates. Although the discussions in Chapters 1 and 2 briefly touched upon the probabilities associated with discontinuous variates such as the proportion of heads in a group of tossed coins and the proportions of black and white balls drawn from a group in an urn, little more was done than demonstrate the use of Pascal's triangle.

Obviously, this primitive technique cannot be used very efficiently except for the simplest kinds of problems. Now the theory of probability and the mechanics of its application will be discussed further. Figure 9-1 is a tree diagram, which illustrates the *multiplication law* of probability. Referring to this diagram, suppose that we start at the point, 0, with a large number of particles. Half go in the direction *A*, and half go in the direction *B*. Now, if at the next fork in the pathways, half of each group turns left to the paths marked *A'*, and the remainder turn to the right to the paths marked *B'*, 1/4 of the original number will be on each of the four paths. If the same process is repeated at the next division of the pathways, 1/8 of the original number will arrive at each of the terminals.

For students of biology or medicine, particularly those interested in physiology, the abstractions depicted by the tree diagrams find concrete illustration in the divisions of the main arterial outflow of the heart. Figure 9-2 shows a diagrammatic representation of the heart and a few major outflow paths. As an anatomical cartoon this representation is merely a schematic diagram and not to be taken as a faithful reproduction. Although this model departs from reality, the principles illustrated are completely valid. Similarly, the fractions about to be used bear about the same relationship to true physiological

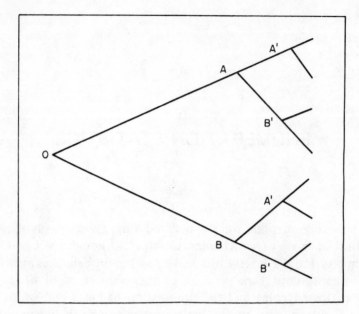

Figure 9-1. Tree diagram illustrating multiplication rule.

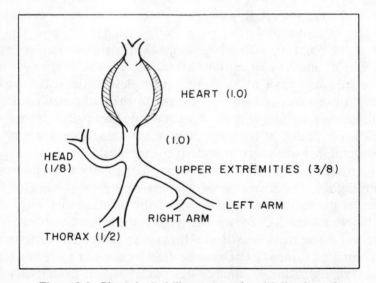

Figure 9-2. Physiological illustration of multiplication rule.

facts, but they will nevertheless help to make the rules to be developed clear and obvious.

As the heart pumps blood into the outflow tracts, the entire output must pass though the aorta. We may say then that the proportion of the blood pumped out of the heart into the aorta is 1.0, the entire amount. This blood will be divided among the major branches. If we were to consider the fate of a single corpuscle starting in the chamber of the heart, we could summarize its possibilities by stating that the corpuscle could pass into the vessels to the head *or* to the upper extremities *or* to the thorax and other structures below. If the corpuscle were to end up, for instance, in the thumb, it would pass first through the main branch to the upper extremities *and* the branch to the right arm *and* to the final branch to the thumb. Note the use of the word *or* in describing the total possible choices and the use of the word *and* in describing a specific path. The importance of the use of these two words in subsequent mathematical considerations will be made clear below.

Suppose now that the fractions of the total output going to the labeled branches are as indicated by the fractions shown in the parentheses in Figure 9-2. When the corpuscle arrives at the first set of alternative paths, the probability that it will go to the head is $1/8$; the probability that it will go to the thorax and lower structures is $1/2$, and the probability that it will go to the upper extremities is $3/8$. Since these account for all of the possible choices, their sum must equal 1.0. Note that when the choices were described, each choice being linked by the word *or*, the sum of the possible choices equaled one. It is useful to remember that when the probabilities are linked by the word *or*, they are added. This is occasionally referred to as the *addition rule*.

If the corpuscle has gone down the path to the upper extremities, it may join $1/2$ of its companion cells going to the left arm or the $1/2$ going to the right arm. In order to obtain the probability that the corpuscle will go down the branch to the upper extremities *and* down the branch to the left arm, we *multiply* the associated probabilities. That is, the probability that it will go into the branch for the upper extremities, $3/8$, is multiplied by the probability of the next choice, $1/2$, so that the final probability that a corpuscle starting from the heart will go to the left arm is $(1/2)(3/8)$ or $3/16$. The operative word here is *and*. When we wish to compute the probability of a succession of choices all linked by the word, *and*, we multiply the associated probabilities. This is known as the *multiplication rule* for compound probability.

To illustrate these rules further, consider a population made up of individuals who may be short or tall and fat or thin. Suppose 3/4 of the group are short, and 1/4 are tall; 2/3 are fat, and 1/3 are thin. If we ask what proportion of the group is tall *and* fat, we apply the multiplication rule and have $(1/4)(2/3) = 1/6$. Of course, if we ask what proportion of the group is short or tall, we are including all of the possibilities, $3/4 + 1/4 = 1.0$. Although this may seem too obvious to mention, such information can be quite useful, as shall be seen in the following example.

Suppose we ask what proportion of the population would be included if we took only those that are short *or* thin. Notice carefully that these are not mutually exclusive categories. The selection here will include individuals who are short and thin as well as individuals who are short and *not* thin. A little reflection accompanied by the constructive use of a little scratch paper will reveal how dangerous it is to use rules blindly. If we attempt to find the answer simply by adding 3/4 and 1/3 to obtain 13/12, we know that the answer is not correct as it is greater than 1.0. We can find the correct answer by noticing that we have excluded the group that is tall *and* fat. This proportion is, as calculated above, 1/6. If this is the group excluded, the remainder must be $1.0 - 1/6$ or 5/6.

We could have arrived at the same answer, using the addition and multiplication rules, by defining more clearly the groups included when we classified those that are short or thin. This classification would include all of the short people regardless of whether they are fat or thin (i.e., 3/4). Then we include any tall people that are thin (i.e., tall *and* thin), which would be $(1/4)(1/3) = 1/12$. Adding 1/12 to 3/4, we would obtain 5/6.

The lesson to be learned from this example is that the addition rule can be applied *only* to mutually exclusive events or conditions. Thus, for example, if we ask what proportion of the group are tall *or* thin, we must note that being tall does not exclude the possibility of being thin (i.e., these conditions are not mutually exclusive).

A brief example may serve to clarify this principle further. Suppose that we are considering a similar group, which is divided into a tall group comprising 1/4 of the total, a medium height group comprising another 1/4 of the total, and a short group making up the remaining 1/2. The ratio of fat to thin is again 2/3 to 1/3. If we ask what proportion are tall *or* short *and* fat, we would find this as (1/4 + 1/2) (2/3) = 1/2. Note that this is the same as excluding from the total or 1.0 the proportion that is medium height *and* fat as well as all of those that are thin. This would yield the same answer calculated as 1.0 − (1/3 + 2/12) = 1/2. This double method of calculating the answer can serve as a valuable check, as it is easy to make mistakes by leaving out groups that are included by implication.

Another useful way to avoid such mistakes is to use the following notation, which is also important in the development of the subject at hand. In the last example, the entire population being considered was made up of those who are tall, those who are of medium height, and those who are short. Mathematically this can be stated as 1.0 (1/4 + 1/4 + 1/2). We can also make the same statement algebraically by substituting symbols for numbers: $1.0 = (T + M + S)$. As far as the attribute of weight is concerned, the entire population may be regarded as being made up of those who are either fat or thin. Thus, 1.0 = (2/3 + 1/3), or stated algebraically, $1.0 = (f + t)$. Since (1.0) (1.0) = 1.0, we may make the following substitution: $(T + M + S)$ $(f + t) = 1.0$.

Carrying out the details of the indicated multiplication, we have $Tf + Tt + Mf + Mt + Sf + St = 1.0$. There are two important consequences of this last maneuver. First, we have defined all of the possible classed based on the attributes under consideration. In this example we have six possible classes. The second important consequence can readily be appreciated when we substitute the assigned values for their algebraic equivalents. For the above example we find the following comparison:

$$
\begin{array}{ccccccc}
Tf & + & Tt & + & Mf & + & Mt \\
(1/4)(2/3) & + & (1/4)(1/3) & + & (1/4)(2/3) & + & (1/4)(1/3)
\end{array}
$$

$$
\begin{array}{ccc}
& Sf & + & St \\
+ & (1/2)(2/3) & + & (1/2)(2/3)
\end{array}
$$

The foregoing discussion is intended to impress upon the student not only the addition and multiplication rules of probability but also the basic notion that we can use symbols for all of the possibilities and, using the ordinary rules of algebra, manipulate either the symbols or the numerical equivalents of the proportions or probabilities that they represent to find the probabilities of various combinations of attributes. This elementary principle is important in the logical development of the statistical analyses based on the expansion of the binomial.

We may begin this development by considering again the population in which 1/4 of the members are tall. Without further classification of the remainder of the population, we categorize them simply into one remaining group, those that are not tall. This exemplifies the general approach to the classification of populations to be analyzed into two groups: (1) those that have the attribute represented by the proportion, p, and (2) those that do not have the attribute represented by the proportion, q. In the present example, the proportion that are tall, p, is 1/4. Thus, $p = 1/4$, and $q = 3/4$, and, as in all other cases, $p + q = 1.0$.

Now if we select a person at random from this population, the probability that we will have selected a tall person is 1/4, while the probability that we will have selected a person who is not tall is 3/4. Suppose now that we select a sample of two people from the population. Although the situation is not very complicated, it is important to understand the steps in the analysis of the possible results. The probability that the first person of the pair selected is tall may be expressed as p_1 and the probability of his being "not tall" as q_1. The probability that the first *and* second were tall would be $p_1 p_2$. The probability that the first was tall and the second was not tall would be $p_1 q_2$. Note that the probability that chance will take one path or the other is strictly comparable to the probabilities associated with branching in the tree diagram or the circulation example (Figures 9-1 and 9-2). Thus, we can use the symbolic approach to summarize the possibilities as $(p_1 + q_1)(p_2 + q_2)$, but, since we are not concerned with the order in which the tall or "not tall" subjects arrived in the sample, we can summarize more simply as $(p + q)(p + q) = p^2 + 2pq + q^2$. As has been demonstrated, we can substitute the equivalent numerical probabilities to obtain 1/16, 6/16, and 9/16, the proportion of the time or probability that we will obtain 2, 1, or 0 tall men in a sample. The same reasoning can be applied to the probabilities for various sample compositions when the sample size is increased to 3. The proportions representing the possible choices for the first person selected, the second, and the third are summarized as $(p_1 +$

q_1), ($p_2 + q_2$), and ($p_3 + q_3$). Then, since we are not concerned particularly with the order in which the specimens arrive in the sample but only with the final composition, the final summary of possible combinations may be represented by $(p + q)^3$.

In general, then, if p is the probability of the occurrence of an event, and q is the probability of its not occurring, the relative frequency in n trials is shown by the terms arising from the expansion of the binomial, $(p + q)^n$. Thus, for example, imagine that we have a large urn filled with spheres that are physically identical except that 1/5 are black and 4/5 are white. If we designate the probability of selecting a black sphere as p and the probability of selecting a white one as q, the binomial expression becomes $(p + q)$ = (1/5 + 4/5).

Suppose now that we take a random sample of 3 spheres. This sample might contain 0, 1, 2, or 3 black spheres. The probability or relative frequency of each of these possibilities is shown by the terms in the expansion of the binomial, $(p + q)^3$.

$$
\begin{array}{ccccc}
p^3 & + & 3p^2q & + & 3pq^2 & + & q^2 \\
f: & (1/5)^3 & + & 3(1/5)^2(4/5) & + & 3(1/5)(4/5)^2 & + & (4/5)^3 \\
& 0.008 & + & 0.096 & + & 0.384 & + & 0.512 \\
x: & 3 & & 2 & & 1 & & 0
\end{array}
$$

Shown here are the algebraic terms in the expansion on the first line. The second line shows the numerical substitution for the algebraic terms with their decimal equivalents below. Note that the sum of these quantities is equal to 1.0. The last line shows x, the number of black spheres in the samples.

Polynomials are analyzed by grouping the terms so as to form a binomial. The following simple example demonstrates this process. Suppose that in an urn as above we have white, black, and yellow spheres in the proportions, 1/8, 3/8, and 4/8. If we take at random a sample of two, the number of yellow spheres in the sample might be 0, 1, or 2. To determine the relative frequencies the essential elements of the problem determine our procedure.

In this instance we are concerned only with the relative frequencies of yellow spheres alone. As demonstrated earlier in this chapter the relative frequencies of any of the various combinations can be expressed and manipulated in terms of their algebraic equivalents. Thus, if we use w, b, and y to express the proportions of each type, we have $(w + b + y)^n = 1.0$ where n is the sample size. When the polynomial contains only three components and when n is small, the complete expansion without regrouping is relatively simple. Thus:

$$(w + b + y)^2 = w^2 + 2wb + b^2 + 2wy + y^2 + 2by$$

Substituting 1/8, 3/8, and 4/8 for w, b, and y would yield the relative frequencies of the various combinations implied by each of the terms. Genetic studies depend on matings, which represent random samplings of two at a time (i.e., $n = 2$). The polynomials representing gene frequencies can be expanded as illustrated above and managed without too much difficulty.

More complicated polynomial samplings are common in such clinical measurements as the differential white blood cell count. Routinely at least five cell types are identified and counted in a random sample of 100 white blood cells. This presents the complexity of five terms in the polynomial and the large n of 100. Resolution of the large n problem is demonstrated later in this chapter. At this point we demonstrate the regrouping of the polynomial as a binomial.

For the problem at hand we regard all of the spheres as either yellow or not yellow. We now rewrite the polynomial in the form of a binomial and carry out the expansion as follows:

$$[y + (w + b)]^2 = y^2 + 2y(w + b) + (w + b)^2$$
$$f\!: \ (1/2 + 1/2)^2 \quad = 1/4 + \quad 1/2 \quad + 1/4$$
$$x\!: \qquad\qquad\qquad 2 \qquad\quad 1 \qquad\quad 0$$

If we use a, b, c, d, and e to represent the proportions of the different kinds of white cells in the circulation, we might similarly divide these into 2 groups as, for example, a and not a. To simplify the expression we would probably let $p = a$, and $q = (b + c + d + e)$.

Notice that we are concerned here with frequencies of an occurrence, x, in a sample of size n. The number, x, is a discontinuous variate obtained by *counting* rather than by *measuring*. The binomial expansion in one form or another serves to analyze this variety of problems. We use the binomial frequency distributions generated by the expansion of $(p + q)^n$. When n is 2, 3, or a little more, the expansion is not very difficult, but as n increases, this soon becomes impractical. In an earlier chapter we showed how coefficients for the expansion could be found using Pascal's triangle, but this becomes unwieldly when n goes much above 10. The best way to find each term of the expansion is to use the point binomial:

$$\frac{n!p^x q^{(n-x)}}{n!(n-x)!}$$

The point binomial determines the relative frequency of exactly x in a sample of n. Thus, for example, if we wanted to calculate the probability of exactly 3 fives (i.e., no more and no less) appearing out of 30 tosses of a die, this would be:

$$\frac{30!(1/6)^3(5/6)^{27}}{3!27!} = 4060(1/6)^3(5/6)^{27} = 0.137$$

We use the binomial frequency distributions to solve the following types of problems:

1. When the proportion, p, is known, what is the probability of x occurrences in a sample of n?
2. When a sample shows the proportion, x/n, what are the confidence limits of our estimate of p, the proportion in the sampled population?
3. When two samples with proportions, x_1/n_1 and x_2/n_2, are obtained, what is the probability of the observed difference occurring by chance if both samples are from the same population?
4. When several samples show different proportions, what is the probability of these differences' occurring by chance if all of the samples are drawn from the same population?

An image of the frequency distributions associated with different ranges of n and p values helps clarify the logic of the various approaches used in solving these problems. Suppose, for example, that in a certain population half the people have Type O blood and that we select 6 of these people at random (i.e., $p = 1/2$, $n = 6$). If we took a great many such samples, we know that on the average we would find 3 with Type O blood, but we also know that these samples of 6 would have anywhere from 0 to 6 Type O individuals. The relative frequencies of the different samples are calculated from the expansion of the binomial, $(1/2 + 1/2)^6$, as tabulated below:

x		f
0	q^6	1/64
1	$6pq^5$	6/64
2	$15p^2q^4$	15/64
3	$20p^3q^3$	20/64
4	$15p^4q^2$	15/64
5	$6p^3q^2$	6/64
6	p^6	1/64

By way of contrast suppose that 1/6 of this population have Type B blood. If again we take many random samples of 6, the relative frequencies will be different. Here n is still 6, but now $p = 1/6$, and $q = 5/6$. The tabulation of the expansion of $(1/6 + 5/6)^6$ follows, but because of the unwieldly numbers (e.g., $q^6 = 15625/46656$), the relative frequencies will be expressed as decimal fractions.

x	f
0	.335
1	.402
2	.201
3	.054
4	.008
5	.001
6	.00...

Figure 9-3 shows a frequency polygon for these two samples. When $p = 1/2$, the distribution is symmetrical around the mean, *3*, but when $p = 1/6$, note how the distribution is skewed. This illustrates the generalization that the more p deviates from 1/2, the more skewed will be the distribution. However, if we increase n, even though p deviated from 1/2, the frequency polygon becomes more symmetrical around the mean, $\bar{x} = np$. Figure 9-4 shows the frequency polygons for $n = 6$ and $n = 18$ when $p = 1/6$. A smooth curve is drawn

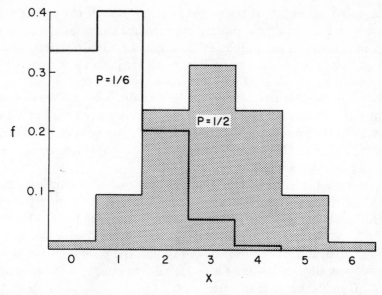

Figure 9-3. Relative frequencies, f, of x for $p = 1/6$ and $p = 1/2$ when $n = 6$.

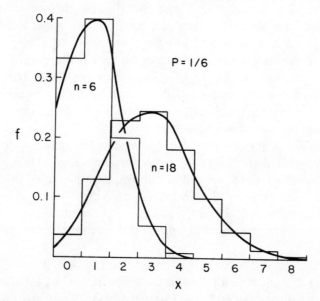

Figure 9-4.   Relative frequencies, f, of x for $n = 6$ and $n = 18$ when $p = 1/6$. Note how a smooth distribution curve drawn through the midpoints approaches a normal distribution as n increases.

through the midpoints at the height of each rectangle to emphasize the contrast and to show how rapidly even modest increases in n cause the frequency polygon to approach the normal distribution curve. Notice also that since the relative or proportionate frequencies are plotted as a function of x, the areas of both polygons are equal to 1.0.

With the relative frequency calculations above we can make useful probability estimates. For example, suppose we need to find a Type B donor in the population described above. If we selected 6 potential donors at random, about 66.5% of the time we would find one *or more* with the type we wanted.

Notice that we wanted to find *at least* one; therefore our need would be met even if there were more than one B donor among the 6. Our probability of success is thus the sum of the probabilities for 1, 2, 3, 4, 5, and 6. We could also determine the probability of finding none, and since this probability plus the probability of one or more covers all possibilities, the probability of none, 0.335, subtracted from 1.0 yields the same answer.

Note carefully that we are not concerned with the probability of finding exactly one Type B (no more and no less) in our sample. We are concerned with finding at least one. When p is known or given, the exact relative frequency (no more, no less) for any given value of x in samples of size n is determined by the point binomial. The proportion of samples showing as many as or more than x is the sum of the point binomials up to $x = n$. The proportion showing as few as x or less is the sum of point binomials down to $x = 0$. Just as in the frequency distributions of continuous variates, these total probabilities correspond to areas under the distribution curve.

For samples of modest size all of the relative frequencies can be more or less easily calculated. As n increases, however, the labor becomes excessive, but as we have pointed out above, as n increases, the polygon approaches the shape of the normal distribution. The mean of this distribution, as expected, will always be np. It can be shown mathematically that the variance is equal to npq, and therefore the standard deviation is $\sqrt{npq}$.

Suppose now that we wish to determine the probability of finding 5 or more B donors in a randomly selected sample of 60 drawn from the population in which Type B occurs in 1/6. Since this probability will correspond to the area under one end of the distribution curve, we use the table of Z values. We calculate Z as follows:

$$Z = \frac{np - x}{\sqrt{npq}} = \frac{60(1/6) - 5}{\sqrt{60(1/6)(5/6)}}$$

Note that the tabulated Z value will correspond to the probability of finding 5 or less. Subtracting this from 1.0 yields the probability desired.

It is generally stipulated that n be equal to 30 or more when the normal distribution is used in place of the frequency polygon, but if p is near $1/2$, the U and Z tables can be used for samples as low as 15.

Ordinarily there is no need to expand large n binomials, but unusual problems may require partial or even complete expansion. Carrying out 25 or more separate solutions to the point binomial would be correct but laborious. The same results are more easily obtained if the process is carried out as a sequence of multiplications. As shown in the example below, it is convenient to calculate the integers and the fractional components separately in a table where each entry corresponds to a specific value of x.

The first column lists the sequence of x values beginning either with n or 0. The first integer or coefficient is always 1, and the second is always n. Multiplying this coefficient by $(n - 1)/2$ yields the third entry. The fourth entry is obtained by multiplying the third by $(n - 2)/3$. Notice that each succeeding entry is obtained by decreasing the numerator by 1 and increasing the denominator by 1. With almost any small hand calculator, all 25 or more coefficients can be tabulated in a few minutes.

If the column of x values is headed by 0, the first fractional entry must be q^n. If the column is headed by n, the first fractional entry must be p^n. In the example shown below for the expansion of $(p + q)^n = (1/3 + 2/3)^{30}$, because $(1/3)^{30}$ is negligibly small, the tabulation begins with $x = 0$ with the corresponding fractional entry of $q^n = (2/3)^{30} = 5.2148(10)^{-6}$. Depending on the availability and capacity of a hand calculator, finding this first fractional entry is the only step in the process that may be troublesome. Each succeeding fractional entry is obtained by multiplying the preceding entry by p/q when the first entry is q^n. When the first entry is p^n, multiply by q/p. As demonstrated below, it is useful to carry out the final multiplication in two steps. Each coefficient is first multiplied by $(p/q)^x$. Multiplying each of these by a constant, q^n, produces the final entries. The tabulation here is carried out only for the first four entries.

$$(p + q)^n = (1/3 + 2/3)^{30}$$

x	coeff.	$(p/q)^x$	product	pt. binomial
0	1	1	1	5.21×10^{-6}
1	30	1/2	15	78.2×10^{-6}
2	$30(29/2) = 435$	1/4	108.8	5.67×10^{-4}
3	$425(28/3) = 4060$	1/8	507.5	2.65×10^{-3}
4	$4060(28/4) = 27405$	1/16	1712.8	8.93×10^{-3}

From the preceding discussions it can now be appreciated that there are four elements in almost any problem involving discontinuous variates or counting. For consistency, in the developments that follow, these will be symbolized as x, n, p, and P. Individuals in a sample showing some attribute are counted as x. The total number in the sample is n. The fraction of the parent population showing the attribute is p, and P is the relative frequency of x in an infinite number of samples of size n. The relationship of these four variables is expressed in the point binomial equation, and, in principle, when any three components are known or implied, the fourth can be determined.

The problems and examples considered above provided n and p explicitly. Given x or some range of x, we calculated P, the probability (i.e., the proportion of an infinite number of samples). A more important and much more common problem involves estimating p, the population proportion, from the sample proportion, x/n, when P is implied by the required level of confidence.

Suppose, for example, that within some large community the proportion of individuals with Type B blood is not known. Obviously a random sample of a few thousand would probably provide a reasonably good estimate of p. This would be costly and time consuming, but without any idea of how large or how small p might be, we would have no basis for estimating the cost of an adequate survey. As demonstrated here, relatively modest "pilot studies" provide a range of such estimates. The larger the sample, the narrower will be the range.

Imagine, then, that we have taken a random sample of 10 individuals and have found that 2 are Type B (i.e., $n = 10$, $x = 2$). At the moment our best estimate of p is 0.2. We realize, however, that p could be either much lower or much higher than 0.2, even if we assume that an incidence of 2 in 10 is not particularly unusual. We assume that whatever the true value of p might be, 2 in 10 could occur by chance as much or more than 5% of the time.

There are two possibilities, p_1 and p_2, where p_1 is the lower of the two. If the true proportion were p_1, a sample of 10 would contain as many as 2 or more, 2.5% of the time. If the true proportion were p_2, a sample of 10 would contain as few as 2 or less, 2.5% of the time. Thus, if we knew p_2 and calculated the point binomials for 0, 1, and 2, their sum would be 0.025, as indicated here:

$$q^{10} + 10q^9 p + 45q^8 p^2 = 0.025$$

The summation used to calculate p_1 would be:

$$1 - [q^{10} + 10q^9 p] = 0.025$$

In principle, each of these equations containing a single unknown could be solved, but in fact, to do so would be difficult and time consuming. Fortunately, tables comparable to the U and Z tables related to the normal distribution have been calculated for the binomial distributions. Table A6-I is used here, and for $x/n = 2/10$, we find $p_1 = 0.025$, and $p_2 = 0.56$. Figure 9-5 shows the frequency distribution polygons for $(p + q)^{10}$ for these two values of p.

The limits in Table A6-I are essentially areas of double tail segments of the binomial histogram. These are comparable to the corresponding segments of the normal curve in Table A2-I for each value of U. Table A6-II corresponds to single tail areas comparable to those tabulated in Table A2-II. The following example demonstrates the use of Table A6-II.

Suppose that tissue fragments isolated from a certain tumor grow as cell cultures successfully 33% of the time (i.e., $p = .33$). An experimenter needs at least 3 vaible cultures for his study. He wants to know how many explants he must make to be 95% certain that he will have at least 3. This implies solving for n in the following equation.

$$0.67^n + n(0.67)^{(n-1)}(0.33) + n!(0.67)^{(n-2)}(0.33)^2 = 0.05$$

With much time an effort this could be solved, but it is more practical to use Table A6-II as follows. The columns of entries are scanned to find for $x = 3$ an upper limit of p close to 0.33. In the columns for $n = 21$, the upper limit is 0.33, and therefore we

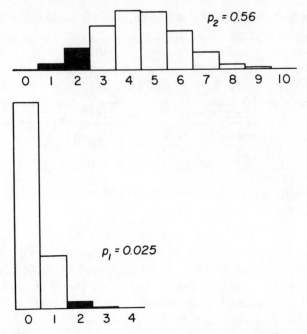

Figure 9-5. Frequency polygons for $p = 0.025$ and $p = 0.56$. Shaded area of polygon for $p = 0.025$ corresponds to proportion of time that x will equal 2 or more. Shaded area on polygon for $p = 0.56$ corresponds to proportion of time that x will equal 2 or less. In both instances, the shaded areas occupy 2.5% of the total area.

recommend preparing 21 cultures to be 95% certain that 3 will be viable.

Notice that we are in effect using this table in reverse. The entries are essentially similar to those in Table A6-I, except that here we find that for an incidence of x in a sample of n the upper and lower 90% confidence limits of p are tabulated. Thus, if we did not know p to be 0.33 and had obtained 3 viable cultures out of 21, the table would indicate that this could have happened by change 5% of the time if p were as high as 0.33.

As noted earlier the binomial distribution begins to approximate the normal curve when n is more than 30. Problems similar to those above are easily solved when treating the binomial as a normal distribution with the mean at np and with a standard deviation equal to $\sqrt{npq}$.

SUMMARY

This chapter deals with attributes that are counted rather than measured. Generally, the proportion of the population that shows some attribute of interest is designated as p; the proportion that does not is designated as q. Therefore $(p + q) = 1.0$.

When a random sample consists of n items, the relative frequency of the attribute occurring x times in samples of this size is expressed by the terms arising from the expansion of the binomial, $(p + q)^n$. A plot of the relative frequencies as a function of x forms a distribution polygon whose mean, $\bar{x}$, is at np.

The probability, P, of exactly x, no more and no less, in a sample size n is expressed by the point binomial:

$$P = \frac{n! \, p^x q^{(n - x)}}{(n - x)! n!}$$

The probability of obtaining x or more is the sum of point binomials for $x, x + 1, x + 2, \cdots, n$. The probability of obtaining x or less is the sum of point binomials for $x, x - 1, x - 2, \cdots, 0$.

When n is small and p deviates from $1/2$, the distribution is skewed, but as n increases, the distribution becomes symmetrical around the mean, np, and approaches the normal distribution with a standard deviation of $\sqrt{npq}$.

When the frequency of an attribute is x in a sample of n, the 95% confidence limits of p are provided in Table A6-I for samples with n less than 30. When n is greater than 30, the normal distribution is adequate, and p is calculated from the relationship:

$$U = 2 = \pm(np - x)/\sqrt{npq}$$

PROBLEMS

1. Approximately 10% of the U.S. black population carry the gene for sickle cell anemia. In this population what is the probability that both partners in a marriage carry this gene?
   What is the probability that only one of the partners carries the gene?
2. Out of 100 randomly selected subjects 15 are found whose serum contains no antibodies to rubella. Estimate with 95% confidence what proportion of the population is susceptible to rubella.
3. Out of seven seagulls in a trap, one is found with detectable traces of DDT in its liver. What are the 95% confidence limits of the seagull population that might be affected?
4. Ten wild mice are trapped. What is the chance probability that this sample will have only two males?

10
THE POISSON DISTRIBUTION

As shown by the French mathematician, S. D. Poisson, an interesting and useful simplification can be made in the point binomial when p is very small and n very large. Under these conditions we can write:

$$\frac{n!p^x q^{(n-x)}}{x!(n-x)!} = \frac{(np)^x e^{-np}}{x!}$$

Both the binomial and the Poisson distributions have their mean at np, but the point binomial requires both the frequency of occurrence, x, and the frequency of nonoccurrence, $(n - x)$. The point Poisson requires only the frequency, x. Thus, for example, we have records that tell us the number of fatal accidents per day, but we have no way of knowing how many accidents did not occur. To use the point Poisson and summations of the Poisson distribution, all we need know is the mean, np. Thus, if we know the mean accident rate per day, the relative frequency of any given daily accident rate can be calculated with the above formula.

The inseparable combination of n and p in transforming the equation of the point binomial to the point Poisson seems to diminish the number of variables. With binomial expansions sample size and number of trials are usually factors determining n. With the Poisson distribution the number of trials or replications may be implied rather than explicitly indicated in the formula. This makes it necessary to play close attention to dimensions and units expressed or implied. The following examples will illustrate the nature of this problem.

Suppose that the relative concentration of some compound being monitored with a radioactive isotope yields 4 counts per minute (cpm). If the observation is limited to only 1 minute, the count recorded could easily be as low as 1 or as high as 8, more than 5.0%

of the time. On the other hand, if counts are recorded for 60 minutes, 95% of the time the total will fall somewhere between 209 and 271. Dividing by 60 to convert these data to counts per minute will again yield 4 cpm as the mean, but now the probable range of estimates will be close to 3 to 5. Note that in the first instance the possible range depended on units of counts per minute, whereas for the range in the second instance the units were counts per hour, which were then divided by 60 to express the range in counts per minute.

The kinds of problems encountered with the binomial distribution occur again with the Poisson. The four variables, n, p, x, and P appear in the Poisson as (np), x, and P, but, as noted in the above example, N, the number of trials may also be important.

When the mean, (np), is known, the relative frequency, P, can be calculated for any value of x. For large values of (np) the normal distribution closely matches the Poisson, and the tabulated U and Z values provide reliable estimates. Because n is very high, even though p is low, when np is greater than 9, the distribution becomes fairly close to the normal.

For smaller values of np enough of the distribution can be quickly calculated to provide the needed data. Consider first the statistical prediction of possible values of x as illustrated in the following example.

Suppose that a small community depends for medical care on the hospital in a larger center some miles away. For emergency care this is unsatisfactory. Therefore, a small unit is to be planned that will provide immediate hospitalization needs for the first 24 hours. Patients requiring additional care would then be transferred to the central hospital. During the preceding year there were 214 emergencies. The planners are anxious to know how many beds the unit should contain. They would like to be reasonably sure (i.e., 95% confident) that enough beds will always be available. The analysis proceeds as follows.

Although 214 cases were recorded for the past year, the number of cases per year might vary. To estimate an annual rate unlikely to be exceeded more than 5% of the time we take $214 + 1.645\sqrt{214}$ or 238 cases per year. The mean number per 24 hours, np, is then $238/365 = 0.652$. The probability of x cases on any given day is calculated with the point Poisson:

$$P = \frac{(0.652)^x e^{-0.652}}{x!}$$

To expedite the generation of a table of P values needed to solve this problem we carry out a sequence of multiplications as illustrated here. Under the heading, x, the possible daily numbers of patients are entered from 0 down to about 5 or 6. Entered in the next column are the corresponding coefficients generated as follows. Corresponding to 0 and 1 the coefficients are always 1 and np, or in this case, 1 and 0.652. Each of the succeeding coefficients is generated in sequence by multiplying the preceding coefficient by np (0.652 here) and then dividing by the corresponding x. We have the coefficient for $x = 1$. To find the next coefficient, we multiply by np and divide by 2. For the following coefficient we multiply again by np and now divide by 3, and continue this process as far as seems useful, usually when the sum of P is very nearly 1.0.

X	coeff.	P	$f/yr.$	ΣP
0	1	.521	190	.521
1	.652	.340	124	.861
2	.213	.111	41	.972
3	.046	.023	9	.995
4	.0075	.004	1	.999

With large values of n the binomial approaches the normal distribution. By definition, a Poisson distribution involves very large values of n. Therefore, except where p is so remote from 1/2 that np is less than 10, the Poisson is close to a normal distribution.

For the Poisson as in the normal distribution the variance is equal to npq, but here p is so small that $q = (1 - p)$ is virtually equal to 1.0. Thus, the mean and variance of the Poisson distribution are both equal to np. These facts are used in the following ways.

Suppose that a 1 gallon sample drawn from a pond contains 10 tadpoles. This is the sample concentration from which we would like to estimate the 95% confidence limits of the mean number of tadpoles per gallon, np. We use the following relationship:

$$\frac{np - x}{\sqrt{np}} = \frac{np - 10}{\sqrt{np}} = 2$$

Solving the resulting quadratic, we find the solutions for np of 5 and 19 tadpoles/gallon.

Next, suppose that in a 10 gallon sample, we had found 100 tadpoles. Notice that in this instance, even though the number of tadpoles/gallon is still 10, the actual number corresponding to x here is 100 tadpoles/10 gallons. Now if we solve for np using the same equation, the mean will be expressed in units of tadpoles/10 gallons.

Our estimate of 95% confidence limits will be 82 to 122. Converting this to units of tadpoles/gallon, we have 8 to 12. These examples emphasize two points. The larger the actual number of items counted, the closer this number approximates the mean. The error expressed as a percentage for a count, x, may be quickly approximated as $200/\sqrt{x}$.

When only 10 tadpoles were counted, the approximate error in estimating np would be $200/\sqrt{10} = \pm 63\%$. When 100 tadpoles were counted, the error is close to $200/\sqrt{100} = \pm 20\%$. We must emphasize here the distinction to be made between the calculated number in some unit and the number actually counted. Thus, for example, in estimating the number of erythrocytes per cubic millimeter in a blood sample, a very small volume of blood is prepared in a dilution of $1/50,000$ (i.e., $p = 0.00002$). If in 1.0 mm^3, 100 cells are counted, the original concentration is estimated as 5,000 cells/mm^3. The possible error is estimated as $200/\sqrt{100} = 20\%$ and *not* $200/\sqrt{5,000,000}$.

SUMMARY

When p is very small and n is very large, the point binomial can be transformed to the point Poisson:

$$P = \frac{(np)^x e^{-np}}{x!}$$

When a mean frequency, np, is 10 or larger, the Poisson distribution begins to approach the normal distribution with the mean and standard deviation equal to np. Relative frequencies and confidence limits are then estimated in the usual way with tabulated values for U and Z.

PROBLEMS

1. An isotope sample in a gamma counter registers 20 counts in 10 minutes. What are the 95% confidence limits of the average count per minute?
2. An isotope sample registers on the average four counts per minute. What proportion of the time will there be two counts or less per minute?
3. Out of 100 white cells examined, 2% are basophils. What proportion of the time can this be expected to happen if the true proportion is 0.9%?
4. The surgical unit of a hospital reports an average of two postoperative wound infections per month. What is the probability of four infections occurring in a month if nothing has changed?

11
THE CHI-SQUARE DISTRIBUTION
AND VARIANCE RATIOS

One of the most useful distributions encountered in theoretical and applied statistics is the χ^2 distribution. This distribution will play a large role in the methods to be developed in the following chapters. The χ^2 distribution arises from a consideration of the function, U, described in Chapter 4. It will be recalled that this function has a normal distribution with mean equal to zero and standard deviation equal to unity. Thus, any variable, x, can be transformed into a comparable value of U through the following relationship:

$$U = \pm\frac{\mu - x}{\sigma}$$

If we were to take a series of measurements, $x_1, x_2, x_3, \cdots, x_n$, we could calculate a corresponding series of values, $U_1, U_2, U_3, \cdots, U_n$. The sum of these approaches zero, but if we square before adding, the sum of the squared U's is known as χ^2 (pronounced chi square). Each sample U is a deviation from its mean at zero. Therefore, the sum of these squared deviations, χ^2, is $(n - 1)$ times the sample variance of U. Thus:

$$\chi^2 = \frac{\Sigma(\mu - x)^2}{\sigma^2} = (n - 1)V_u$$

The relationship of χ^2 to a sample variance can be shown by expanding the squared term and carrying out the summation as follows:

$$\sigma^2\chi^2 = n\mu^2 - 2\mu\Sigma x + \Sigma x^2 \tag{1}$$

A sample variance can be expressed as:

$$V = \frac{\Sigma(\overline{x} - x)^2}{(n - 1)} \qquad (2)$$

Expanding the squared term, summating, and transposing $(n - 1)$, we obtain the following:

$$(n - 1)V = n\overline{x}^2 - 2\overline{x}\Sigma x + \Sigma x^2 \qquad (3)$$

Using equations (1) and (3) to solve for Σx^2, we obtain, after the usual algebraic manipulation, the following relationship:

$$\sigma^2 \chi^2 = (n - 1)V - n(\overline{x}^2 - \mu^2) + 2\Sigma x(\overline{x} - \mu) \qquad (4)$$

When samples are large, or when we consider $\overline{x}$ as the average of the means of many small samples as implied in the next step, the values for $(\overline{x}^2 - \mu^2)$ and $(\overline{x} - \mu)$ are negligibly small, so that the last equation can be simplified and rewritten as:

$$\chi^2 = (n - 1)\frac{V}{\sigma^2} \qquad (5)$$

Now if we let n be the value of any sample size greater than one, we can take many samples of this same size. The resulting χ^2 values from these samples would then form a frequency distribution. Notice that in this case the only value that would vary with each sample, assuming that all were drawn from the same population, would be the value for V. Thus, although the value of n is fixed, the frequency of χ^2 values for this sample size would vary. Figure 11-1 shows the shape of the curve obtained for 9 degrees of freedom.

One use of the χ^2 distribution may now be demonstrated. Suppose that we have made a series of 10 measurements and have calculated a standard deviation, s, from the observed variance, V. Now we would like to know the 95% confidence limits for s. First we obtain the confidence limits for V as follows. Table A-7 gives the χ^2 values corresponding to $P = 97.5\%$ and $P = 2.5\%$ for 9 degrees of freedom. The nature of this procedure is revealed by considering the χ^2 distribution curve illustrated in Figure 11-1. Unlike the normal

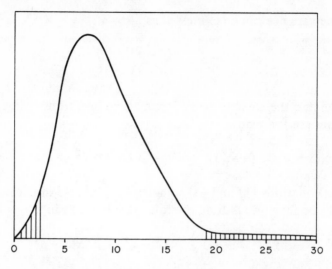

Figure 11-1. χ^2 distribution showing a total of 5% of population in the shaded tails (2.5% under each tail).

distribution curve, this curve is far from symmetrical. Since we are seeking the limits consistent with 95% of the values that can appear from an infinite number of random samples, we need to chop off 2.5% of the population from either extreme as shown by the shaded areas at the two ends of the distribution curve. As indicated in the figure, the χ^2 values are 2.70 and 19.0, respectively. To use these values in obtaining the upper and lower confidence limits, rewrite equation 5 as in the following formulas where V_1 and V_2 are lower and upper limits, respectively, and $\chi_1{}^2$ and $\chi_2{}^2$ are the χ^2 values that were just obtained:

$$V_1 = (n-1)\frac{V}{\chi_1{}^2} \qquad V_2 = (n-1)\frac{V}{\chi_1{}^2}$$

The potential uses of these algebraic abstractions may be more readily understood by considering an oversimplified example. For this purpose, let us assume that although the mean height of men is larger than the mean height of women, the variances are nearly identical. Then, if we took random samples from either sex, the variances of each of these samples would probably differ from the universe variance, σ^2. Nevertheless, we could predict that 95% of the time the ratios of sample variance to universe variance, V/σ^2, would fall within certain limits. The random distribution of such ratios make up the χ^2 frequency distribution curve. A ratio falling outside the limits cutting off 21/2% from each end of the curve would be consistent with the notion that factors other than random chance were present.

Notice that up to this point we are considering random samples made up entirely either of men or women. That is to say, the samples are homogeneous with respect to the attributes under consideration. The essential point to grasp now is that if we took random mixed samples made up of both men *and* women, far fewer than 95% of the V/σ^2 ratios would fall within the previously determined χ^2 limits. Thus, it is clear that this ratio and the limiting χ^2 limits tabulated for corresponding degrees of freedom can be used to test the homogeneity of a sample when σ^2 is known.

The application of these principles can be illustrated as follows. Suppose that there are four groups of samples. The numbers in each sample need not be the same. These groups may be people or things from four different settings, patients with the same disease treated in four different ways, or botanical specimens from four different regions. In each group the incidence of some attribute varies, and we wish to determine if this variation is attributable to chance or possibly to the differing circumstances.

Tabulating the data with the groups labeled as A, B, C, and D, as shown in Table 11-1, the next columns show first the number in each group and next the number showing the attribute. We see that out of 160 specimens, 40 show the attribute in question. We assume that if each of the groups is a random sample from a homogeneous population, 40/160 or 1/4 of each group may be expected to show the attribute. Thus, $p = 1/4$ and $q = 3/4$. If n is the number in each group, the expected number having the attribute will be np, and if x is the number that actually have it, we can express the observed frequency of each group in terms of $U = \pm(np - x)/\sqrt{npq}$. As shown in the table the sum of $U^2 = \chi^2$.

In this instance $\chi^2 = 16.08$. Note carefully that the degrees of freedom here are $4 - 1 = 3$ and are *not* $160 - 1 = 159$. Although 160 specimens provide the data, these result in only 4 estimates of U. Table A-7 shows that with 3 degrees of freedom χ^2 exceeds 12.8 by chance less than 0.5% of the time. We conclude then that there is a significant difference between the groups.

Table 11-1

GROUP	No. of Specimens	Observed no. with Attribute	Expected no.	
	(n)	(O)	(E)	(U^2)
A	21	12	5	12.44
B	64	10	16	3.0
C	28	8	7	0.19
D	47	10	12	0.45
SUM:	160	40	40	16.08

To determine which of these groups is most likely to represent a different population, we consider the individual estimates of U^2 as χ^2 with one degree of freedom. This will exceed 3.84 by chance less than 5% of the time. Noting that group A shows a value of 12.4, we are justified in concluding that this group is significantly different from the others in this sampling.

Probably the most frequently used application of the χ^2 distribution involves the comparison of two proportions, x_1/n_1 and x_2/n_2. A typical example would be n_1 in an experimental or treated group with x_1 of these showing some attribute or effect and n_2 in a control or comparison group. Usually the data are summarized in a 2 × 2 contingency table, and the subsequent calculations of χ^2 are so easy that potential pitfalls and misapplications often occur without notice. These will be explained after the essentials of this application are illustrated in the following example.

Suppose for example that out of 20 cancer patients receiving a new treatment modification 9 survive beyond 5 years. Out of 81 patients receiving the unmodified treatment 29 are still alive after 5 years. The data are customarily summarized as shown in Table 11-2. Assuming the null hypothesis we look at the combined totals and see that 29/101 survive regardless of treatment received. Therefore, out of a group of 20 we would expect 5.7 to survive 5 years and 14.3 to be dead. As soon as a single expected value is calculated (e.g., 20 × 29/101) = 5.7), all of the remaining expected values as shown in the parentheses are easily found by subtraction from the marginal totals.

Table 11-2

	New Treatment	Old Treatment	Total
Alive	9 (5.7)	20 (23.3)	29
Dead	11 (14.3)	61 (57.7)	72
Total	20	81	101

Note that in each box there is now an expected number and an observed number. In the upper left box the expected number is $E = 5.7$ and the observed number is $O = 9$. χ^2 is calculated from these entries as the sum of $(E - O)^2/E$ as follows:

$$(5.7 - 9)^2/5.7 + (23.2 - 20)^2 + (14.3 - 11)^2 + (57.7 - 61)^2$$
$$= (3.2)^2(1/5.7 + 1/23.3 + 1/14.3 + 1/57.7) = 3.12$$

There is only one degree of freedom here even though there seem to be four calculations of $(E - O)^2/E$. This apparent paradox will be explained more completely below. For the moment, however, it should be noted that once the marginal totals are in place, as soon as any number is selected for one of the boxes, the remaining three boxes can be filled in only one way.

These are the mechanics of using the χ^2 distribution to compare two proportions. There are certain limitations that must be remembered before this simple analysis can be expected to yield reliable estimates. First, if n is less than 30 or if any of the calculated expected values are less than 5, the exact probability calculation appropriate for small samples must be used. This exact calculation is a little more difficult to understand and a little more laborious. It is discussed in Chapter 12.

The term, $(E - O)^2/E$, is in some applications a manifestation of a Poisson distribution, but in the 2 × 2 contingency table as demonstrated above, it derives from the binomial distribution. This will become clear when we consider the first sample in the pair of proportions compared above. For the data in the upper left box $E_1 = n_1 p$, and $O_1 = x_1$. Then,

$$(E_1 - O)^2/E_1 = (n_1 p - x_1)^2/n_1 p$$

This refers to the patients in the first group that survive. The box just below refers to the patients that do not survive. E_2 in this case is $n_1 q$ or $n_1(1 - p)$, and we now write:

$$(E_2 - O_2)^2/E_2 = [n_1(1 - p) - (n_1 - x_2)]^2/n_1 q$$
$$= (n_1 p - x_1)^2/n_1 q$$

When these are added, the sum from the first 2 boxes becomes:

$$\frac{(n_1 p - x_1)^2}{n_1 p} + \frac{(n_1 p - x_1)^2}{n_1 q} = \frac{(n_1 p - x_1)^2}{n_1 pq} = U_1{}^2$$

Similarly, data from the remaining two boxes yields $U_2{}^2$, and then, $U_1{}^2 + U_2{}^2 = \chi^2$. Since only two deviations are squared, χ^2 has only one degree of freedom. By remembering that P is less than 0.05 when χ^2 with one df is greater than $(1.96)^2 = 3.84$, these comparisons can be evaluated without the use of tables.

When n is fairly large, comparison of 2 proportions may be accomplished as described above without further modification. The χ^2 distribution is useable here, because as n becomes large, the polygon representing the terms of the binomial expansion approach the contours of the normal distribution curve. The frequency polygon divides the smooth curve into narrower segments as n increases and eventually become negligibly small. When n is less than 50, the relatively coarser divisions produce a significant difference between the area under the polygon and under the curve. Since the curve passes through the center of each of the rectangles making up the polygon, the correct point of contact with the curve is made by decreasing each deviation from the mean, $\pm(np - x)$, by 1/2 unit. Thus, in the example above, the deviation, $(3.2)^2$, should be changed to $(2.7)^2$. In this instance the conclusion is not changed, but to avoid erroneous conclusions, it is prudent always to decrease the deviations by 1/2 in calculating χ^2 for 2 $\times$ 2 contingency arrays. This alteration is known as Yate's correction.

It is important to remember that comparisons of proportions in a 2 $\times$ 2 contingenty table with the χ^2 distribution are valid only when the samples and frequencies are large enough to treat the binomial as a normal distribution. Therefore, when n is less than 30 and when any of the E values is less than 5, one must use the exact method discussed in the chapter on comparison of small sample proportions.

The first part of this chapter showed how the χ^2 ratio of a sample variance, V, to the universe variance, σ^2, is described by the χ^2 distribution for the appropriate degrees of freedom. Finding confidence limits of the σ^2 from sample V is comparable to finding confidence limits of μ from a sample mean, $\bar{x}$, except that in the latter, σ^2 is assumed to be known. When σ^2 is unknown, estimation of confidence limits of μ from $\bar{x}$ requires the Student's t distribution. Similarly, when two variances are to be compared and σ^2 is unknown, the F distribution of Snedecor is used.

If a great many pairs of samples are drawn from the same populaion, the ratio of the larger variance of the pair to the smaller will generate a frequency distribution curve extending asymmetrically around a mean of 1.0. These distributions will vary with the degrees of freedom in each of the samples. Mathematically derived equations for these distributions provide tables of probabilities associated with these ratios (see Table A-8).

When two variances are to be compared, we test the null hypothesis assumption that both represent the same population. The probabilities of ratios as large or larger than critical values of F are found in the table.

In general, to test the hypothesis that V_1 with $(n_1 - 1)$ df and V_2 with $(n_2 - 1)$ df arose by chance from the same population, we calculate F as V_1/V_2. The larger variance is always in the numerator. The critical F ratios for probabilities of 1, 5, and 10% are tabulated for the degrees of freedom of each variance.

As a specific example, suppose that the variance calculated from 10 measurements is 7, and we wish to compare this with a variance of 21 calculated from 5 measurements. The F ratio is always written with the degrees of freedom as subscripts, the df of the larger first. In this instance we write $F_{4,9} = 21/7 = 3$. In Table A-8 we find that the chance probability of this ratio is less than 10 but more than 5%. The F ratio here would have to be larger than 3.63 to be regarded as having a chance probability of less than 5%.

Although the F test has obvious application in comparing the precision of two methods of measurement, it plays its most important role in the analysis of variance discussed in a later chapter.

SUMMARY

The χ^2 distributions are the frequency distributions of the ratio of sample variances to the square of the universe standard deviations.

Thus, $\chi^2 = (n-1)V/\sigma^2$. Since these distributions are known and tabulated, the confidence limits of universe variance can be estimated from a sample variance. The χ^2 for the level of confidence and degrees of freedom, as obtained from the tables, is used to estimate the corresponding confidence limits using the relationship: $\sigma = \sqrt{(n-1)V/\chi^2}$.

The χ^2 distribution may also be used to estimate the probability of obtaining by chance the deviations between expected and observed values of both continuous and discontinuous variables. In general $\chi^2 = \Sigma U^2$ or $\Sigma(\mu-x)^2/\sigma^2$ for continuous variables and $\Sigma(np-x)^2/npq$ for discontinuous variables. When n is greater than 30 and each expected value is greater than 5, proportions may be compared in a 2 × 2 contingency table, which yields a χ^2 with one degree of freedom.

If two random samples are selected from the same normally distributed population, the estimates of the variance based on these samples may be expected to vary. The ratios of pairs of variances are necessarily a function of the size of each sample and form the corresponding F distributions. The F test consists of finding the ratio of the larger variance of a pair to the smaller. This is compared to critical values for different levels of probability.

PROBLEMS

1. Find the 95% confidence limits for the following estimated standard deviations:
   a. $SD = 10, n = 5$ b. $SD = 10, n = 20$ c. $SD = 10, n = 30$
2. In one part of an encampment with 50 men, sanitation rules were rigidly enforced, but during a winter epidemic 30 at one time or another came down with upper respiratory infections. In another part of the encampment sanitation rules were observed in the usual way. Out of 100 men, a total of 85 became ill during the winter.
   Compare the χ^2 values with and without Yate's correction. Do the data suggest that rigid enforcement of rules was effective?
3. Out of 40 students at school A, 10 succeeded in finding summer employment. Only 5 out of 30 from school B were successful. Does this prove that school A students were more enterprising? Compare χ^2 values with and without Yate's correction.
4. Two methods for blood oxygen measurements were being compared. A sample measured five times with method A yielded a standard deviation of 10% of the mean. With method B a sample measured eight times yielded a standard deviation of 8% of the mean.
   a. How would you express the F ratio?
   b. What is the probability of obtaining these results by change if there is no difference between the methods?

12
COMPARING PROPORTIONS IN
SMALL SAMPLES

Samples large enough to use the χ^2 calculation described in the previous chapter are not always available, and even when samples are substantial, one of the entries in the contingency table may be less than five. Large samples are obviously preferable to small samples, but time and expense are not the only factors that may compel us to examine small samples. Accidents involving a limited number of cases may provide data impossible to duplicate or extend.

The analysis of small samples requires a more complicated, carefully reasoned use of binomial expansions as demonstrated in the following example. Suppose, for example, that out of 4 patients receiving treatment A only 1 recovers, whereas out of 8 comparable patients receiving treatment B, 6 recover. We would summarize the results in a contingency table as follows:

	A	B	Total
Alive	1	6	7
Dead	3	2	5
Total	4	8	12

Treatment B appears to be better than treatment A, but before accepting this conclusion, we must consider the probability that these results could occur by chance if there were no difference between the treatments. We assume, therefore, that out of the 12 patients 7 were destined to recover regardless of whether they received treatment A or B. The assignment to one or the other treatment was then a matter of chance.

A useful statistical model at this point is a collection of 12 physically identical balls except that 7 are white and 5 are black. Repeatedly these could be separated at random into a pair of groups, A with 4 and B with 8. Group A could have anywhere from 0 to 4 white balls, and the corresponding number of white balls in group B would be 7 minus the number in A. The possible combinations are limited to the following:

A	0	1	2	3	4
B	7	6	5	4	3

To calculate the probability of observing a difference in sample proportions as large or larger than $6/8 - 1/4$, we calculate the compound probability of A's containing one or less and B's containing 6 or more.

There is more than one way to arrive at this calculation, but for reasons detailed later in this chapter, we regard the following approach as the most useful. We assume that the mixture of black and white balls making up this sample are drawn from an infinite population in which the proportion of white balls is p and the proportion of black balls is q. For samples of size n the probability of any number of white balls from 0 to n in the sample is shown by expansion of the binomial, $(p + q)^n$.

When we draw paired samples of 4 and of 8, we are, in effect, drawing a sample of 12. The number of white balls in such a sample can vary from 0 to 12, and in each instance these will be randomly distributed either to subgroup A or subgroup B. Thus, if we took an infinite number of samples of size 12, we would at the same time be taking an infinite number of samples of 4 and of 8. We are concerned, however, only with those samples of 12 that contain 7 white balls. This proportion, P, is expressed by the point binomial as:

$$P = \frac{12p^7 q^5}{7!5!} = 792\,p^7 q^5$$

In each of the instances included in this proportion the white and black balls are randomly divided between subsets A and B.

Consider now the probability that A will contain, for example, one white ball. This would be expressed by the point binomial:

$$\frac{4!pq^3}{3!} = 4\,pq^3$$

Only that fraction of this proportion would be part of P above, when at the same time B contained 6 white balls. This probability would be expressed as:

$$\frac{8! \, p^6 q^2}{6!2!} = 28 \, p^6 q^2$$

The probability that A will contain 1 white ball and that B will contain 6 white balls is expressed by the compound probability:

$$(4 \, pq^3)(28 \, p^6 q^2) = 112 \, p^7 q^5$$

Group A can have anywhere from 0 to 4 white balls, but for each of these instances only that fraction will form part of P when group B contains a corresponding number to make a total of 7 white balls as indicated above. Thus, we can tabulate all of the compound probabilities associated with these pairs as follows:

A	B	$Pr(A)$	$Pr(B)$	$cpd\ Pr$
0	7	q^4	$8\,p^7 q$	$8\,p^7 q^5$
1	6	$4\,pq^3$	$28\,p^6 q^2$	$112\,p^7 q^5$
2	5	$6\,p^2 q^2$	$56\,p^5 q^3$	$336\,p^7 q^5$
3	4	$4\,p^3 q$	$70\,p^4 q^4$	$280\,p^7 q^5$
4	3	p^4	$56\,p^3 q^4$	$56\,p^7 q^5$

$$\Sigma = 792 \, p^7 q^5 = P$$

As expected, the sum of these compound probabilities is equal to the point binomial expressing the probability of finding 7 white balls in a random sample of 12. We are concerned, however, only with those instances where A is one or less with corresponding values for B of 6 and 7. From the tabulation above we see that these occur with relative frequencies of $8\,p^7 q^5$ and $112\,p^7 q^5$. The sum forms a proportion of the total equal to the chance probability that proportions found in samples of 4 and 8 will differ by as much or more than $6/8 - 1/4$. Here we find this to be:

$$\frac{120\,p^7 q^5}{792\,p^7 q^5} = 0.15$$

We have used the samplings of white and black balls as a statistical model to test the null hypothesis with regard to the results of using treatments A and B on two small groups of comparable patients. Since the same results could have occurred by chance 15% of the time if there were no difference between A and B, the data is insufficient to establish the superiority of treatment B.

Now that we have demonstrated the analysis of a specific example, we may summarize in general terms the evaluation of differences in proportions in small samples. In this summary the proportions will be designated as A $= x_1/n_1$ and B $= x_2/n_2$, where x_1/n_1 is the smaller of the two. Some of the expressions will be simplified if we let $x_1 + x_2 = X$, and $n_1 + n_2 = N$. A contingency table would summarize the data as follows:

	A	B	Total
(+)	x_1	x_2	X
(−)	$(n_1 - x_1)$	$(n_2 - x_2)$	$(N - X)$
Total	n_1	n_2	N

The point binomial providing the denominator for the final probability calculation is then:

$$P = \frac{N!}{X!(N - X)} p^X q^{(N - X)}$$

As indicated below the computations are facilitated by forming five columns, which for this description we have designated with Roman numerals. Columns I and II list the pairs of numbers that add up to X. Column I represents the possible values for sample A beginning with 0 and progressing to x_1. Column II representing B must then start with X and decrease with each step down to x_2. Column III lists the point binomials corresponding to the numbers in A. The general form is shown below using a to represent the successive numbers in column I. Column IV lists the point binomial expressions corresponding to the numbers in column II. The general form shown here uses b to represent the successive numbers in column II. The product of each of the pairs forming columns III and IV are listed in column V. Note that if no mistakes have been made, the exponents on p and q in this column will always be the same as those in the formula for P. The sum of these products indicated as $\Sigma(\text{cpd.Pr})^*$

---

*Σ(cpd. Pr) = sum of compound probabilities.

is the numerator of the final calculation. Using P as calculated above for the denominator yields the probability of obtaining by chance a difference in sample proportions as large or larger than $(x_2/n_2 - x_1/n_1)$.

I	II	III	IV	V
A	B	Pr(A)	Pr(B)	cpdPr
0	X	$\dfrac{n_1!\,p^a q^{(n_1-a)}}{a!(n_1-a)!}$	$\dfrac{n_2 p^b q^{(n_1-b)}}{b!(n_2-b)!}$	$\dfrac{n_1!\,n_2!\;p^X q^{(N-X)}}{a!(n_1-a)!\,b!(n_2-b)!}$
·	·			
·	·			
·	·			
·	·			
x_1	x_2			$\overline{}$
				$\Sigma(\text{cpd. Pr})$

$$\Pr(x_2/n_2 - x_1/n_1) = \frac{\Sigma(\text{cpd. Pr})}{P}$$

There appears to be a striking contrast between methods used for evaluation of differences in proportions between small samples and those used for large samples. A relationship does exist between these differing approaches that is not immediately obvious, but that can be revealed by the following considerations.

In the small sample procedure outlined above, the point binomial, P, included *all* of the possible combinations of A and B. Thus, a sharply defined and limited universe of discourse is established. Figure 12-1 shows the relative frequencies of the possible combinations that make up this universe. Note that the shaded areas representing the relative frequencies of the A and B combinations, (0,7) and (1,6), occupy 15% of the total area.

A similar histogram can be formed for any pair of proportions being compared. Note, however, that the number of divisions on the horizontal axis varies with X, the sum of x_1 and x_2. As the sample size increases, the histogram more closely approximates the normal distribution curve. When the sample size and the numbers involved become large enough to make binomial calculations excessive, the approximation to the normal curve is reasonably close. The observed proportions, x_1/n_1 and x_2/n_2, may then be regarded as deviations from the mean X/N, as described in the χ^2 calculation.

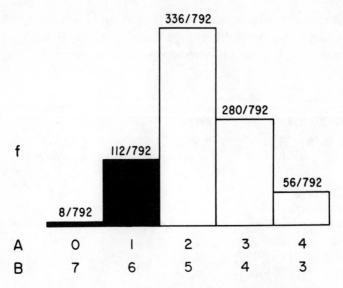

Figure 12-1. Random samples of 12 balls are drawn from an unknown mixture of black and white balls. Out of those instances in which 7 are white, f is the proportion of these instances in which A white balls are in the subset of 4, and the remainder, B, in the subset of 8.

PROBLEM

A proposed modification for improving the treatment of a malignant disease is being compared with the standard method at a certain clinic. Five years after the start of the comparisons, three out of seven (i.e., 42.9%) receiving the standard treatment are dead, but only one out of the eight (i.e., 12.5%) receiving the modified treatment had died. What is the probability of obtaining these results by chance if there is no real improvement?

13
ANALYSIS OF VARIANCE

Most experiments or investigations involve the comparison of one group with another. In previous chapters we have shown how the t test or some similar criterion is used to estimate the probability that the two groups considered are samples drawn from the same parent population. There are times, however, when more than two alternatives must be considered. It may be necessary to evaluate three or more alternative treatments, analytic methods, or other possible courses of action. Using the previously described methods, we might try to compare all of the possible combinations as pairs. With three groups labeled A, B, and C, we could compare A with B, A with C, and B with C. There are now three differences to consider, whereas previously there was only one.

The probability is 0.05 that two randomly selected sample means from the same population will differ by more than t times the standard error. But if we have three randomly selected means, the probability is 0.14 that one of the differences will exceed this limit. With more than three comparisons this probability becomes larger, and thus increases the risk of seeing differences where none exist.

To estimate the probability that differences among three or more sample means could occur by chance, we use a process known as *analysis of variance*. Because this process requires several calculations on each of the data sets, the procedure seems complicated. If the procedures are carried out mechanically without a clear understanding of their rationale, one is apt to make serious errors. On the other hand, a thorough understanding of each of the components helps prevent these errors and provides a basis for more effective design of experiments or collections of data.

Essentially, analysis of variance rests on the relationship of the variance of a population to the variance of means of samples drawn from this same population. We used this relationship in Chapter 5 when we pointed out that $\sigma_{\bar{x}} = \sigma_x/\sqrt{n}$ or $n\sigma_{\bar{x}}^2 = \sigma_x^2$. It follows, then, that if we have the means of several groups whose sample size is n, we can use the variance of these means, V_m, for one estimate of the population variance, i.e., $nV_m = V_1$. We can obtain a second estimate of the population variance, V_2, by calculating the variances within each sample and using the mean of these estimates. We can then use the F test to estimate the probability that the variances were drawn from the same population.

These are the bare essentials, but in order to make full use of them we must first consider the relationship of the variance of some measurement to the variances of its components. Suppose, for example, that we are estimating the variance of the distribution of weights in a given population of specimens. From the data obtained we could calculate the observed variance, and if the precision of our weighing instrument were perfect, this observed variance would be the variance of the population in question.

Suppose then that our weighing instrument is less than perfect so that part of the variation in the observed weights reflects the variable behavior of the instrument. We could test this idea easily enough by weighing one specimen repeatedly. If the calibration of the scale were correct, the mean of the population of observations on the weight of a single specimen would coincide with its correct weight. The variance observed in these repeated weighings would be the variance of the instrument and would have been incorporated in the variance of weights observed on the population of specimens. Our problem, then, is to find a way to correct our data for the variable behavior of the weighing instrument. Analysis of variance involves the resolution of variances into their components, and therefore we will use this example in deriving some general rules about variance.

In these derivations it will be simpler and clearer if we replace the cumbersome elementary standard deviation relationship, $s = \sqrt{\Sigma(\bar{x} - x)^2/(n - 1)}$, with expressions of variance in terms of the sum of squares. If the reader will sharpen his pencil, gather a good supply of scratch paper, and fill in the details where needed, the derivations below will be easy to follow and to understand.

The deviation of each variable, x, from its mean, $\bar{x}$, is squared, and the sum of these squares is $\Sigma(\bar{x} - x)^2$. We can summarize this process as follows:

$$(\bar{x} - x_1)^2 = \bar{x}^2 - 2\bar{x}x_1 + x_1{}^2$$
$$(\bar{x} - x_2)^2 = \bar{x}^2 - 2\bar{x}x_2{}^- + x_2{}^2$$
$$\bullet$$
$$\bullet$$
$$\bullet$$
$$\bullet$$
$$\frac{(\bar{x} - x_n)^2 = \bar{x}^2 - 2\bar{x}x_n + x_n{}^2}{\Sigma(\bar{x} - x)^2 = n\bar{x}^2 - 2\bar{x}\Sigma x + \Sigma x^2}$$

We define $\bar{x}$ as $\Sigma x/n$, and therefore by substitution the summation can be written as:

$$\Sigma(\bar{x} - x)^2 = \Sigma x^2 - (\Sigma x)^2/n, \text{ or } \Sigma x^2 - n\bar{x}^2$$

Both of these forms will be used, but the one most convenient for the derivation under consideration will be chosen.

Returning to our example of weighings, we can regard the observed weight, z, as being made up of two components: x, the true weight, and y, the deviation introduced by the instrument. Thus:

$$z = x + y$$

Suppose, as suggested above, we weigh the same individual r times and obtain a series of results, $z_1, z_2, \cdots, z_r$. The mean of these results will be $\bar{z} = \Sigma z/r$. The details are summarized as:

$$z_1 = x + y_1$$
$$z_2 = x + y_2$$
$$\bullet$$
$$\bullet$$
$$\bullet$$
$$\frac{z_r = x + y_r}{\Sigma z = rx + \Sigma y}$$
$$\bar{z} = \Sigma z/r = (rx + \Sigma y)/r = x + \Sigma y/r$$

Since y may represent either a positive or negative error, $\Sigma y/r$ tends to become negligibly small as we increase the number of replicate weighings.

These data are used to calculate the variance of the instrument from the following relationships:

$$(r - 1)V_z = \Sigma z^2 - r\bar{z}^2$$

Proceeding as demonstrated above, we can demonstrate that:

$$\Sigma z^2 = rx^2 + 2x\Sigma y + \Sigma y^2$$
$$r\bar{z}^2 = rx^2 + 2x\Sigma y + (\Sigma y)^2/r$$
$$(r - 1)V_z = \Sigma x^2 - r\bar{z}^2 = \qquad\qquad \Sigma y^2 - (\Sigma y)^2/r = (r - 1)V_y$$

We already knew, as demonstrated here, that repeated weighing of the same specimen reveals the variance of the instrument. It is more important to note that when one component of a variable is constant, it has no effect on the variance.

Our survey data include both specimen and instrument variance, and the details of the calculations may be summarized as follows. The individual weights, $z_1, z_2, \cdots, z_r$ will be used to calculate the mean and variance in terms of their x and y components. We will substitute $x_1 + y_1$ for z_1, $x_2 + y_2$ for z_2, and so on. These steps are summarized as follows:

$$(n - 1)V_z = \Sigma z^2 - (\Sigma z)^2/n \qquad\qquad (1)$$
$$\Sigma z = \Sigma x + \Sigma y \qquad\qquad (2a)$$
$$(\Sigma z)^2/n = (\Sigma x)^2/n + 2\Sigma x\Sigma y/n + (\Sigma y)^2/n \qquad\qquad (3b)$$
$$z^2 = x^2 + 2xy + y^2 \qquad\qquad (3a)$$
$$\Sigma z^2 = \Sigma x^2 + 2\Sigma(xy) + \Sigma y^2 \qquad\qquad (3b)$$
$$\Sigma z^2 - (\Sigma z)^2/n = [\Sigma x^2 - (\Sigma x)^2/n] +$$
$$[\Sigma y^2 - (\Sigma y)^2/n] + 2[\Sigma(xy) - \Sigma x\Sigma y/n] \qquad\qquad (4)$$

Note that the first two bracketed expressions in equation (4), as the equivalent of $(n - 1)$ times the variance of x and y, can be written as:

$$(n - 1)V_z = (n - 1)(V_x + V_y) + 2[\Sigma(xy) - \Sigma x\Sigma y/n] \qquad (5)$$

Except for the remainders enclosed in the last bracket, these show that the variance of z is equal to the sum of the variances of its components. That this bracketed remainder becomes negligible with large samples can be demonstrated by regarding each of the values of x and y in terms of their relationship to the same means, $\bar{x}$ and $\bar{y}$. Then:

$$x_1y_1 = (\bar{x} + \Delta x_1)(\bar{y} + \Delta y_1) \qquad\qquad (6a)$$

or

$$x_1y_1 = \overline{xy} + \bar{y}\Delta x_1 + \bar{x}\Delta y_1 + \Delta x_1\Delta y_1 \qquad\qquad (6b)$$

This can be written for each of the n measurements and added.

$$\Sigma(xy) = n(\overline{xy}) + \bar{y}\Sigma\Delta x + \bar{x}\Sigma\Delta y + \Sigma(\Delta x \Delta y) \qquad (7)$$

Then, since $n(\overline{xy}) = \Sigma x \Sigma y/n$,

$$[\Sigma(xy) - \Sigma xy/n] = \Sigma(\Delta x \Delta y) \qquad (8)$$

In any moderately large series the deviations, Δx and Δy, will be positive about half the time and negative about half the time, and the sums of the positive and negative products will be nearly equal so that $\Sigma(\Delta x \Delta y)$ is likely to be negligibly small. Thus, when variable, z is made up of two components so that $z = x + y$, its variance is equal to the sum of the variances of the components:

$$V_z = V_x + V_y \qquad (9)$$

Two important points must be noted about this final relationship. First, the variance of z will also be equal to the sum of variances of x and y when $z = x - y$. The reader will find it a useful exercise to repeat the derivation outlined above starting with this relationship.

This would seem to imply that if we had started with $y = z - x$ instead of its exact equivalent, $z = x + y$, we could now prove that $V_y = V_z + V_x$, which would not be consistent with equation (9). The fallacy arises here by violating the requirement that the component variables must vary independently. In the equations, $z = x + y$ or $z = x - y$, x and y vary independently; variations in x are not influenced in any way by variations in y. On the other hand, when we write $y = z - x$, this is algebraically correct, but variations in x here produce variations in z, and therefore z and x cannot be treated as independently varying components. Thus, in our weighing example the deviation in the true weight of a given specimen from the mean has no effect on the deviation of the weight shown by the instrument from the specimen's true weight.

The second point to be emphasized is that the remainder shown in the last bracket of equation (5) may safely be regarded as negligibly small only when the sample size is substantial. With smaller samples there is always likely to be some remainder, and the sum of component variances may differ from the observed total variance.

The application of these derivations can be shown with a simple example which will make it easier to follow the algebraic generalizations. Suppose then that we are comparing the duration of relief from pain by three different analgesics labeled A, B, and C. Three groups of 8 patients are selected at random, and the number of minutes that elapse before recurrence of pain is recorded. Each group is treated with a different analgesic. The results are shown in Table 13-1.

Using the null hypothesis, we assume that there is no difference between the analgesics and that the differing durations of action merely reflect variations between patients. In effect, we assume that these results would have been obtained if 24 patients selected at random had been given the same analgesic and then randomly assigned to three groups. We would then expect that the variance of the means for each group of 8 would be 1/8 of the population variance.

Table 13-1

	A	B	C	row mean
	19	15	26	20
	14	12	22	16
	8	3	22	7.7
	31	28	27	28.7
	18	13	17	16
	31	26	38	31.7
	22	20	21	21
	33	35	35	35
sum	176	152	200	
column mean	22	19	25	
ΣZ^2	4440	3632	5576	
$(\Sigma Z)^2/n$	3872	2888	5000	
$(n-1)v$	568	744	576	
v	81	106	82	$\bar{v} = 89.9$

Source of variation	Sum of squares	df	V
Within columns	$c(r-1)V_c = 1888$	21	89.9
Between columns	$r(c-1)V_{\bar{x}} = 8 \times 18 = 144$	2	72
Total	$(rc-1)V_T = 2032$	23	

$$F_{2,21} = 72/89.5 = 0.8$$

In this example the variance of the means is 9, and the population variance may then be estimated 8 times 9 or 72. It would seem now that we could estimate the population variance simply by pooling all of the data and calculating the resulting variance. As will be evident from the algebraic restatement of these calculations below, a better estimate is obtained using the mean of the variances within the groups. As shown in Table 13-1 this is 89.9. We can test the hypothesis that this variance and the variance from the group means represent the same population of variances using the F test described in the previous chapter.

Ordinarily, the ratio, F, is calculated by dividing the smaller variance into the larger. In this application, however, if there were any significant difference between the means of the duration of drug effect, the variance estimated from the means of the groups should be larger than the inherent variance of the population. Therefore, in this kind of analysis, the variance estimated from the means is always placed in the numerator. In this example the variance estimated from the means is smaller than the mean variance within the columns. Since F values for significant differences are always greater than 1.0, we can state that the data reveal no significant difference in duration of action between the three analgesics tested.

The rationale of the steps outlined above can be appreciated if we use appropriate algebraic expressions in place of the numbers. We may then regard each measurement in general as being made up of three components, μ, x, and e. In this instance μ is the universe mean response, x is the deviation from this mean characterizing the drug tested, and e is the deviation arising from variation in patient response. In other kinds of measurements, hemoglobin levels for example, μ would represent the universe mean, x would represent the individual deviation, and e would represent the experimental error.

We may now examine the process when r measurements or observations are made on c individuals. We use r and c to express the numbers of replications here because the data are summarized in r rows and c columns

I	II		c
$\mu + x_1 + e_I$	$\mu + x_2 + e_{II}$	$\cdots\cdots$	$\mu + x_c + e_{c1}$
$\mu + x_1 + e_{I2}$	$\mu + x_2 + e_{II2}$	$\cdots\cdots$	$\mu + x_c + e_{c2}$
$\vdots$			
$\mu + x_1 + e_{Ir}$	$\mu + x_2 + e_{IIr}$		$\mu + x_c + e_{cr}$

Although each entry above has three components, each represents a single number that is used in subsequent calculations. Therefore, we will let $\mu + x + e = z$. The entries in the first column may then be represented as z_{I1}, z_{I2}, z_{I3}, and so on up to z_{Ir}.

The sum of squares needed to calculate the variance of the data in the first column would be:

$$V_{z_I} = V_{e_I} = \frac{\Sigma e_I^2 - r\bar{e}_I^2}{(r-1)} \tag{10}$$

Notice that since μ and x_1 are unchanging components in each entry of the first column, they have no effect on the variance *within* the first column. V_{e_I} provides us with an estimate of σ_e^2, but we know that a larger sample size would be likely to bring our estimate of V_e closer to σ_e^2. Therefore we estimate the variances within each of the columns and use the mean of these estimates, $\bar{v}_e$, as our best estimate. Since each column's estimate is based on $(r-1)$ degrees of freedom, there will be $c(r-1)$ total degrees of freedom when there are c columns. Thus:

$$\bar{v}_e = \Sigma V_c/c = \left[\frac{(\Sigma e_{II}^2 + \Sigma c_{II}^2 + \cdots \Sigma e_c^2) - r\Sigma\bar{e}^2}{c(r-1)}\right] \tag{11}$$

or

$$\bar{v}_z = \bar{v}_e = \frac{\Sigma e^2 - r\Sigma\bar{e}^2}{c(r-1)} \tag{12}$$

Next we estimate the variance *between* columns calculated as the variance of the column means. Then:

$$V_{\bar{z}} = \frac{\Sigma\bar{z}^2 - (\Sigma\bar{z})^2/c}{(c-1)} = \left[\frac{\Sigma x^2 - (\Sigma x)^2/c}{(c-1)}\right] + \left[\frac{\Sigma\bar{e}^2 - (\Sigma\bar{e})^2/c}{(c-1)}\right]$$
$$+ 2\left[\frac{\Sigma(x\bar{e}) - (\Sigma x)(\Sigma\bar{e})/c}{(c-1)}\right] \tag{13}$$

or

$$V_{\bar{z}} = V_x + V_{\bar{e}} + 2\left[\frac{\Sigma(x\bar{e}) - (\Sigma x)(\Sigma\bar{e})/c}{(c-1)}\right] \tag{14}$$

Except for the remainder enclosed in the last bracket, the variance between column means is made up of the variance of x (in this instance the variance in analgesic effect), and the variance of $\bar{e}$ (in this instance the variance between means of the patient effects in each group). Since there are r items in each column, $V_{\bar{z}} = V_e/r$.

When we use the null hypothesis, we are assuming that there is no difference between the means of the columns, i.e., $x_1 = x_2 = x_3 \cdots = x_c$. If x does not vary between columns, in equation (14) V_x

and the bracketed remainder become equal to zero, and $V_{\bar{z}} = V_{\bar{e}} = V_e/r$. Thus, $rV_{\bar{z}}$, the variance *between* columns provides another estimate of V_e, which we can now compare with the estimate in equation (12), the variance *within* columns. We make this comparison using the ratio for the F test as follows:

$$F = rV_{\bar{z}}/\bar{v}_z \tag{15}$$

$V_{\bar{z}}$ is calculated from the means of c columns and is therefore based on $(c - 1)$ degrees of freedom, while $\bar{v}_z$, as shown above, is based on $c(r - 1)$ degrees of freedom.

It is useful at this point to examine in detail the role of the components of the F ratio used above. The effect of each component will be easier to see if we indicate our estimate of the variance of e between columns as $V' = rV_{\bar{e}}$ and the estimate of this variance within columns as $V'' = \bar{v}_e$. The bracketed remainder will be written as R. Then by substituting the relationships indicated in equations (12) and (14) in equation (15) we have:

$$F = \frac{V'}{V''} + \left[\frac{rV_x + 2r(R)}{V''}\right] \tag{16}$$

As we noted above, if the null hypothesis is true, the quantity bracketed in equation (16) will be zero. Then, since V' and V'' are estimates of the same variance, $\sigma_e{}^2$, F should be close to 1.0. On the other hand, if there are large enough differences between $x_1, x_2, \cdots$, and x_c, so that the ratio is definitely larger than 1.0, the F value will exceed the 5% critical value appropriate for the degrees of freedom of the components. Notice that even if V_x is not very large compared to V', increasing r, the number of measurements in each column, will make its presence evident.

Before we can clearly explain the way analysis of variance is used with an improvement in experimental design, we must consider the relationship of the variances within and between the columns of data to the total variance, that is, the variance we can calculate when all of the data are pooled. With c columns of r items, this total variance, V_T, will have $(rc - 1)$ degrees of freedom. In terms of sums of squares we can write:

$$(rc - 1)V_T = r\left[\Sigma x^2 - \frac{(\Sigma x)^2}{c}\right] + \left[\Sigma e^2 - \frac{(\Sigma e)^2}{rc}\right]$$
$$+ 2\left[r\Sigma(x\bar{e}) - \frac{\Sigma x \Sigma e}{c}\right] \tag{17}$$

and from equations (12) and (13), we can write:

$$c(r-1)\bar{v}_z = \Sigma e^2 - r\Sigma\bar{e}^2 \tag{18}$$

and

$$(c-1)V_{\bar{z}} = \left[\Sigma x^2 - \frac{(\Sigma x)^2}{c}\right] + \left[\Sigma\bar{e}^2 - \frac{(\Sigma\bar{e})^2}{c}\right]$$
$$+ 2\left[\Sigma(x\bar{e}) - \frac{\Sigma x\Sigma\bar{e}}{c}\right] \tag{19}$$

From these relationships, with a moderate amount of tedious algebraic manipulation we can show that:

$$(rc-1)V_T = c(r-1)\bar{v}_z + r(c-1)V_{\bar{z}} \tag{20}$$

Note that in this last relationship each of the components consists of a variance multiplied by degrees of freedom, making them equal to the sums of squares so that we can write:

(Total sum of squares) = (sum of squares within columns)
+ r(sum of squares between columns) (21)

We take advantage of this relationship by summarizing our calculations in this standard conventional form:

	Sum of squares	Degree of freedom	Variance
Within columns	$c(r-1)\bar{v}_z$	$c(r-1)$	$\bar{v}_e$
Between columns	$r(c-1)V_{\bar{z}}$	$(c-1)$	$rV_x + V_e$
Total	$(rc-1)V_T$	$(rc-1)$	

From the analgesic data above our formal summary would be:

	Sum of squares	Degrees of freedom	Variance
Within columns	1888	21	89.9
Between columns	$4(36) = 144$	2	72
Total	2032	23	

Although we need only the variances between and within columns to calculate F, calculating the total sum of squares and using the conventional form above provides a valuable check on our arithmetic. Furthermore, as shown below, the total variance is necessary for the next level of analysis of variance.

The data used in the example above fail to demonstrate any appreciable difference between the analgesics tested. We have used the word "appreciable" here intentionally. We imply here that even if there is some difference between the drugs tested, it is obscured by the much larger differences between patients. In short, the variance, V_e, is so much larger than V_x, that the ordinary deviations of Ve from σ_e^2 may well be larger than σ_x^2. We may, in effect, be trying to detect a whisper against the background noise of a boiler factory.

As suggested earlier, we could decrease the background effect by increasing substantially the number of observations in each of the columns. It is more efficient, however, to decrease the effect of the large variation between individuals by improving the design of our experiment. In the specific example used here we can readily understand that in addition to variation between subjects, there is also likely to be fairly wide differences in the causes of their pains. The effectiveness of one of our analgesics on a headache due to eyestrain must surely differ from its effectiveness on a headache arising from a fractured skull.

Suppose then that instead of randomly selecting 24 subjects to be distributed into three test groups, we select 8 subjects and try each of the three analgesics on each subject. The order in which each analgesic is tested on each subject would be varied in some random way. An important point will be illustrated if we use the same data shown in Table 13-1, but now the numbers in each *row* represent the effect of each analgesic on a *single* patient. The observed effect, z, is now made up of components μ, x, y, and e. The symbols, μ and x have the same meaning as before, but now the effect of the basic cause of the pain is reflected in y, and e is the variable response of patients.

The data array may then be shown here in general terms, but since μ is a constant component of all of the data, it will not be included in any of the expressions below.

$$I$$

$$
\begin{array}{llll}
x_1 + y_1 + e_{I1} & x_2 + y_1 + e_{III1} & \cdots\cdots & x_c + y_1 + e_{c1} \\
x_1 + y_2 + e_{I2} & x_2 + y_2 + e_{III2} & & x_c + y_2 + e_{c2} \\
x_1 + y_3 + e_{I3} & x_2 + y_3 + e_{III3} & & x_c + y_3 + e_{c3}
\end{array}
$$

$$
\begin{array}{llll}
x_1 + y_r + e_{Ir} & x_2 + y_r + e_{IIr} & & x_c + y_r + e_{cr}
\end{array}
$$

The columns' means (without μ) are then $x_1 + \bar{y} + \bar{e}_I$, $x_2 + \bar{y} + \bar{e}_{II}$, . . ., $x_c + \bar{y} + \bar{e}_c$. The row means are $\bar{x} + y_1 + \bar{E}_1$, $\bar{x} + y_2 + \bar{E}_2$, and so on. Note that when we use a Roman numeral subscript with $\bar{e}$, this mean value derives from the vertical summation of e's in the designated column. For easier reading, instead of using the Arabic numerals to indicate summation of e's horizontally, we will use the upper case, E. Thus, $\bar{E}_1, \bar{E}_2, \cdots, \bar{E}_r$ indicate the row means of e.

From equation (14) we are able to summarize the sum of squares between columns as:

$$(c-1)V_{\bar{z}} = (c-1)V_x + (c-1)V_{\bar{e}} + 2\left[\Sigma(x\bar{e}) - \frac{\Sigma x \Sigma \bar{e}}{c}\right] \quad (22)$$

In the same way we could write the sum of squares between rows as:

$$(r-1)V_{\bar{z}} = (r-1)V_y + (r-1)V_{\bar{E}} + 2\left[\Sigma(y\bar{E}) + \frac{\Sigma y \Sigma \bar{E}}{r}\right] \quad (23)$$

With algebraic substitutions as outlined previously the total sum of squares can be expressed as:

$$
\begin{aligned}
(rc-1)V_T = {} & r(c-1)V_c + c(r-1)V_y + (rc-1)V_e \\
& + 2r[\Sigma(x\bar{e}) - \Sigma x \Sigma \bar{e}/c] + 2c[\Sigma(y\bar{E}) + \Sigma y \Sigma \bar{E}/r] \quad (24)
\end{aligned}
$$

The rest of this development can be summarized with the following tabulations. Although we have simplified this presentation by leaving out the bracketed components, the reader can easily demonstrate for himself that the final results are algebraically exact. Notice that multiplying the components of equation (22) by r changes $V_{\bar{e}}$ to an estimate of V_e. Similarly, multiplying equation (23) by c changes $V_{\bar{E}}$ to another estimate of V_e.

	Sum of Squares	df
r(between columns)	$r(c-1)V_x + (c-1)V_e$	$(c-1)$
c(between rows)	$c(r-1)V_y + (r-1)V_e$	$(r-1)$
sum between cols. and rows	$r(c-1)V_x + c(r-1)V_y$ $+ V_e(r+c-2)$	$(r+c-2)$
Total	$r(c-1)V_x + c(r-1)V_y$ $+ (rc-1)V_e$	$(rc-1)$
Residual	$(rc-r-c+1)V_e$	$(r-1)(c-1)$

To demonstrate the application of these developments the analysis of the analgesic data is summarized below. Note that when we subtract the sums of squares between rows and between columns from the total sum of squares, the residual is the error variance, V_e multiplied by the degrees of freedom, $(r-1)(c-1)$.

	Sum of squares	df	Variance
r(Between cols.)	144	2	72
c(Between rows)	1768	7	252.6
sum	1912	9	
Residual	120	14	8.57
Total	2032	23	

$$F_{2,14} = \frac{72}{8.57} = 8.4 \qquad 8.4 > 3.74 \qquad p < 0.05$$

$$F_{7,14} = \frac{252.6}{8.57} = 29 \qquad 29 > 4.28 \qquad p < 0.01$$

Note that in this analysis the error variance estimated from the sum of squares between column means is shown by the F ratio to be significantly greater than the error variance estimated from the residual. In our first analysis, variation in patient response was not separated out from the error variance estimated from sum of squares within columns. Thus, even though the same numerator is used in both analyses, the larger denominator in the first analysis results in an F ratio below the critical level.

That variation between patients has a large effect on the estimated error variance is made even more evident by the large ratio of variance between rows to the residual. These calculations demonstrate how a change in experimental design with appropriate modification of the procedure for analysis of the data enables us to identify two sources of variance in addition to the variance of experimental error. It is of some interest at the same time to note that no difference between the column means would have been demonstrated by the t test.

More elaborate experimental designs can be used to cope with more complex comparisons. Subsets of replications are used as outlined in the above developments. The fundamental principle involved in developing procedures appropriate for additional sources of variance can be grasped from the following consideration. Note that in the analysis using variances between means of columns and means of rows, we were dealing with three unknowns, V_x, V_y, and V_e. The algebraic equation expressing variance between column means, the equation for variance between row means, and the total variance equation provides the necessary number of equations to solve for each of the unknowns. We use the F test to show that quantities expressed by these relationships are unlikely to arise by chance.

PROBLEMS

Each of the problems below presents three sets of measurements. Use an analysis of variance to estimate the probability of each set of three arising by chance from the same population. In your answers use the conventional summarizing form that shows along with degrees of freedom the total sum of squares and the sums of squares within and between the columns of data. Calculate the needed variances and the F ratio.

1. A	B	C		2. A	B	C
1.1	1.2	1.2		4.1	3.2	2.2
1.1	1.1	1.2		4.2	3.1	2.2
1.4	1.4	1.0		4.4	3.4	2.0
1.1	1.2	1.3		4.1	3.2	2.3

14
QUALITY CONTROL

Most biological or scientific investigations that depend on quantitative measurements made over an extended period of time can be improved by using some form of quality control. This is particularly true when the measurements depend on unstable reagents or on equipment subject to wear and tear. The need is even more acute in large clinical laboratories where different technicians perform the same tests and individual variations in technique and skill are superimposed on all other sources of variation.

In the performance of quantitative measurements, we are generally concerned with the day to day monitoring of the accuracy and precision. These two terms have specific meaning in this connection. *Accuracy* refers to the nearness of the measurements obtained to their true value. For a given procedure, accuracy is estimated by performing either physical or chemical measurements on a known standard. Thus, for example, if an investigation depends on a titration with a standardized solution of alkali, we check the reliability of this reagent from time to time by titrating a known or gravimetrically determined quantity of acid salt.

Precision refers to the identity or at least the similarity between two or more measurements of the same quantity. To some degree precision is related to the stability of the experimenter's technique, which may need more improvement than he realizes. However, depending on the nature of a particular measurement under consideration, lack of precision may appear because of faulty temperature control, chipped glassware, or worn, corroded or loose parts in the instruments utilized.

The procedures and statistical techniques used for quality control were originally intended and designed for use on industrial production lines. Some of these methods are readily adapted to the laboratory uses indicated above. Although the heavy emphasis on graphic displays of day to day accumulations of data are useful here as illustrative material, the biological or medical worker can do about as well with an appropriately arranged column of figures.

The evaluation of accuracy is a statistical evaluation of the temporal fluctuations of measurements performed on a known standard. Evaluations of precision are related to estimates of the standard deviation. The following examples illustrate the application of statistical principles to the expeditious solution of these problems.

Suppose we are engaged in an investigation requiring a daily (or possibly a weekly) group of nitrogen determinations such as measuring the rate of nitrogen fixation in plants or the rate of protein changes in the serum of animals. In any event, assume that we are using the classical Kjeldahl method which presents a wealth of opportunities for error. Some errors are easily recognized, but some are insidious. The last step in this procedure involves the estimation of the amount of ammonia taken up in a known quantity of acid by back-titration with standard solution of alkali. In addition to the usual variety of blunders and experimental errors, there may be progressive deterioration of the alkaline solution as carbon dioxide is absorbed from the air. Where the ammonia content of the atmosphere is relatively high (i.e., if the laboratory is near the animal colony or if someone has left the stopper off the bottle of ammonia reagent), the measurements err on the high side.

Notice that the type of error described above is *not* related to the skill or meticulous technique of the investigator. The problem here is to detect evidence of deterioration in the measurement *before* it is sufficiently advanced to interfere with the validity of the data being accumulated. There are several ways to accomplish this, a few of which shall be demonstrated in the ensuing paragraphs.

To control the process outlined above, a solution of ammonium sulfate whose concentration is gravimetrically determined is used as a known standard. A portion of this standard is measured along with each batch of unknown samples. The known concentration of the standard is accepted as the true mean, μ. The concentration of each unknown, as measured each time, is x. In Figure 14-1 the mean level, μ, is represented by the horizontal line across the middle of the control chart. Each time we measure the standard chemically, we

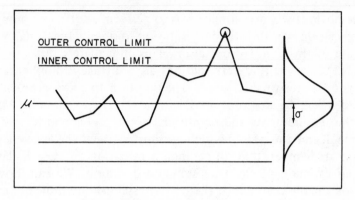

Figure 14-1. Quality control chart.

indicate the result with a dot on the scale, beginning at the left and progressing to the right.

Because of the chance, uncontrollable factors in the measurement, the dots form a random distribution around the central value. The small normal curve at the right side of the chart indicates this distribution. In other words, with this procedure we generate a population of results. If, in addition to knowing the mean, we know the standard deviation, we can determine how far the dots may be expected to deviate from the horizontal line, as long as the process is in control. If the conditions of the measurement change because of factors such as those described above, a different population of results will be generated. When such a change occurs, we apply the null hypothesis. In doing so, we assume that the process is in control. The probability of the deviation of any single result can be estimated when we are reasonably sure of the standard deviation. If the probability is unacceptably low, we reject the null hypothesis and conclude that the process is out of control. With a graph such as the one in Figure 14-1, this change in the population is easily recognized when one or more of the dots appear beyond the parallel boundary lines (at $\pm 3\sigma$ in this illustration). When the process is in control, deviations beyond the boundary as indicated are improbable.

In this example involving ammonia nitrogen determinations, the universe mean, μ, is established by the known concentration of the ammonium sulfate standard. Before the control program can be placed in full operation, we need a satisfactory estimate of σ; then we can proceed to determine the probability level at which the outer boundary lines should be located.

The obvious way to estimate σ is to perform a sufficient number of determinations on the standard solution to obtain a reasonably good estimate of the standard deviation. This immediately raises the questions: (1) how many determinations are a sufficient number, and (2) how good is reasonably good? The solution to these problems and questions will be demonstrated with a specific example illustrating how the requirements and conditions of the problem at hand dictate to some degree the way in which we must arrive at the answers.

First, remember that the standard deviation to be calculated is only an estimate of the true standard deviation. We can, however, with reasonable confidence (i.e., 95%) establish the fiducial limits of this estimate by using the χ^2 distribution, as shown in Chapter 11. Thus, we can state:

$$\sigma = s\sqrt{(n-1)/\chi^2} \tag{1}$$

The true standard deviation may be either larger or smaller than the standard deviation estimated from the available data. Suppose, for example, that we overestimate the standard deviation. If we set the outer control limits at 2 standard deviations as estimated, values falling outside the level of two *true* standard deviations will still fall within the overextended boundaries set by the overestimated values for the standard deviation. The whole purpose of the control chart is to alert us to the possibility of trouble in our method, and with the overestimated values these warnings will be missed. Thus, by using the lower fiducial limit of the standard deviation, we may be alerted to the possibility of trouble more often than necessary, but at the completion of the investigation we will be that much more certain of the validity of the measurements. As indicated below, however, the difficulties arising from too frequent warnings are largely minimized by other considerations in the interpretation of the control chart patterns.

From Eq. (1) for estimating the fiducial limit of the standard deviation, we can see that we should use the χ^2 value that is largest for the degree of probability demanded. Also, if we convert this equation into an expression for the factor multiplied by the estimated standard deviation to convert to its fiducial level, the factors can be tabulated as a function of sample size. This is done for several sample sizes in Table 14-1. Included in this table are the associated factors for setting the control limits at levels equivalent to 2σ and 3σ.

The causes of the variations around the central line are classified as either assignable or unassignable causes. The assignable causes are

Table 14-1

Degrees of Freedom	Factors for Inner Confidence Limit	
	0.05	0.01
1	1	1.31
2	1.14	1.49
3	1.21	1.59
4	1.28	1.67
5	1.31	1.72
6	1.36	1.78
7	1.38	1.81
8	1.41	1.85
9	1.43	1.87
10	1.45	1.90
14	1.48	1.98
19	1.56	2.04
24	1.59	2.06
29	1.61	2.12

the ones which presumably may be indicated or recognized when the control chart indicates that the process is out of control. The unassignable causes are the unrecognized or uncontrollable factors which diminish precision. This latter group often causes variations from one day to another, that exceed the variations encountered during the same day. Thus, if the standard deviation is determined from a replicate series of analyses performed at the same time, the observed standard deviation will probably be smaller than that which would be observed if the series were carried out over several days. This day to day variation appears in the experimental data, and therefore if these data are evaluated on the basis of the more optimistic level obtained during one day's run, they may be misleading. Thus, we need some way to determine the variation from day to day. This does not mean, however, that the standard deviation should be redetermined each day.

Recalling the lessons of Chapter 13, we can see that there is a "within the day" variance and "between days" variance. The variation around the central line within control limits is the sum of these two variances. The analysis in this case can be treated in a simpler, more straightforward manner.

Instead of attempting to obtain an estimate of the standard deviation at the beginning of the study earlier, it is generally more efficient to use the following procedure. On the first day of the study and each day thereafter, the known control sample is measured only once. The unknown samples are measured in duplicate. Of course, if there are many unknown samples each day, this duplication may be limited to a sample size determined with the aid of the factors in Table 13-1. As more data accumulate, the decrease in the number of duplicates required for each day's run will depend on the nature of the procedure involved. Eventually, only three or four duplicates are required.

The duplicate determinations generate a group of differences, d_1, d_2, d_3, $\cdots d_n$. The standard deviation may be calculated with the following formula:

$$s = \sqrt{\Sigma d^2 / (2n)} \qquad (2)$$

where n is the number of *pairs*. The degrees of freedom here are equal to n, *not* $(2n - 1)$. This standard deviation may be used to establish the levels of the control limits parallel to the central line. The variance obtained in this way is the "within the day" variance, V_w. As a sufficient number of days pass increasing the number of determinations on the known control solution, these values are used to determine the total variance, V_t. The variance between days, V_b, is found through the following relationship: $V_t = V_b + V_w$.

Frequently, with certain types of measurements, the F-test will indicate that no significant difference exists between the total variance and the "within the day" variance. The location of the known control data will then suffice to indicate the level of control, and generally the duplication of measurements can be abandoned. On the other hand, if there is a significant difference, a daily correction factor may be derived from the deviation of the known control from the center line. Application of this factor to the unknown data will increase the accuracy of the measurements. Needless to say, these corrections should be made judiciously, if at all. In these cases experience and thorough knowledge of the methods being used are the best guide. It is more important to seek out, wherever possible, the cause of either day to day variation or spuriously close agreement between duplicate determinations.

Once a satisfactory determination of the standard deviation has been made, the control limits may be placed at the equivalent of either 2 or 3 standard deviations from the center line. Two standard deviations, or 1.96, to be more exact, should include 95% of the expected deviations from the center line. Again, depending on an experimenter's judgment, he may regard a point outside this limit optimistically as merely cause for suspicion, or, more pessimistically, as evidence that the set of determinations associated with the deviation are out of control and therefore totally unacceptable. In most applications of quality control charts, deviation beyond 2 standard deviations is regarded as a warning that something *may* be going wrong; deviation beyond 3 standard deviations is taken to mean that something *has* gone wrong. The associated data are then regarded as unacceptable and are discarded.

Obviously, an event of this kind can be disastrous in some experiments. In a sense, this is sounding the alarm when it is already too late to do anything about it. As indicated previously, one of the functions of the control system is to give warning early enough to take corrective action. Thus, it becomes important to consider the patterns which develop so that early clues may not be overlooked.

When a process is in complete control, the dots are more or less evenly distributed on both sides of the center line within the boundaries of the inner control limits (i.e., within 2 standard deviations). Under these circumstances the probability that a given measurement will fall on one side of the center line is $1/2$. If four or five values in succession fall on one side of the center line by chance, the associated probabilities are $1/16$ and $1/32$, respectively. Regarding these as indices of improbable events, we must accept this as evidence that a shift has occurred in the level of the measurements. Falling within the critical boundaries, the data are still acceptable, but this is the time to recheck the reagents and equipment instead of waiting for points to appear beyond the outer critical limit (i.e., beyond 3 standard deviations).

Again, if the process is fully in control, the probability that a given value is higher than the one which preceded it is $1/2$. If a series of values appear in the data in which four or five values in a row are each successively higher, this should be regarded as evidence of a trend. The same reasoning applies, of course, to a similar number of values each successively lower. In either event, the presence of a trend is indicated. Deterioration of reagents is the most common cause, and reagents should be replaced by fresh solutions before the values recorded go beyond the control limits. Occasionally, a trend of this kind may reveal unsuspected wear in parts of the physical equipment or deterioration of electronic components such as vacuum tubes and photocells. These should be checked and corrected as soon as possible when they are the source of uncontrolled variation.

Except for the use of lower fiducial limits for the standard deviation in setting up the inner and outer control limits, the foregoing is a summary of quality control as it is practiced both in laboratories and in industry. The use of quality control systems in biological and medical laboratories is not as widespread as it might be. Aside from the many instances that are attributable to unfamiliarity with quality control methods, most of the resistance to the use of quality control can be blamed on the disproportionate amount of arithmetical labor involved. This is particularly true when measurement processes are not performed often enough to produce a valid estimate of the standard deviation.

Once again we find that a procedure that has long been used intuitively has a sound statistical basis. Whenever we are called upon to make some measurement with a technique that we have not used frequently, we check ourselves by performing the measurement in duplicate. We tend to do this whether the procedure involves a chemical process, such as titration, or a physical process such as weighing. If the two results check fairly closely, we are reassured. If they do not check closely, we do not accept the results and repeat the procedure until a pair of satisfactory results are obtained. The problem of utilizing minimal inspection and as little calculating as possible has been just as important in industrial quality controls as in the biological laboratories. As might be expected, then, a statistically valid evaluation of the procedure outlined above has been made. It can be shown that 95% of the time the mean of a pair of measurements lies within a range of 3.16 times the difference between the pair from the true mean of the population. Thus, if x_1 and x_2 represent a pair of measurements, we may summarize this as follows:

$$\mu = (x_1 + x_2)/2 \pm 3.16(x_1 - x_2) \tag{3}$$

The application of this factor may be illustrated as follows. If we weigh a crucible on an analytical balance with the usual precision so that the two results are within 0.2% of the average, we could express this as:

$$x_1 - x_2 = 0.002(x_1 + x_2)/2 \tag{4}$$

Substituting this in Eq. (3), we find that:

$$\mu = (1 \pm 0.00632)(x_1 + x_2)/2 \tag{5}$$

This would suggest that even with precision of 0.2%, our accuracy may be no better than 0.63%.

Although this simple process is hardly a complete substitute for the entire quality control procedure, it does give a useful method for evaluating the accuracy of measurements that are not performed frequently.

SUMMARY

Quality control is an application of statistical methods for monitoring the accuracy and precision of laboratory measurements. Accuracy is evaluated by measurements on known samples. Precision is a measure of the reproducibility of measurements on the same sample and is estimated from the standard deviation calculated from multiple measurements on the same sample.

When laboratory measurements are made at intervals over an extended period of time, quality control makes the early detection of deterioration in technique, reagents, or equipment possible. The data accumulated as part of a quality control program are important for evaluating the reliability of each measurement as well as for the detection of changes adversely affecting the accuracy of the measurements.

PROBLEMS

1. The columns below give pairs of determinations of hemoglobin measured daily on a large pooled sample of frozen, hemolyzed blood. Prepare a quality control chart showing the best estimate of the mean and the precision of the test on a daily cumulative basis. Compare the variance between days with the variance within days.

x	x'
15.2	15.0
14.8	14.9
15.8	15.4
15.3	14.9
15.0	14.5
15.8	15.6
14.5	14.7
14.6	14.8
14.9	15.1
15.1	15.0
15.2	15.0

15
TESTING ALTERNATIVES TO THE
NULL HYPOTHESIS

The null hypothesis as introduced in Chapter 6 has been used frequently in this text. The perceptive reader, particularly if he has had some experience accumulating experimental data, is likely to suspect that the application of this form of reasoning may not tell the complete story. As we apply the null hypothesis in the comparison of two means, we ask what the probability is of obtaining the observed difference by chance if there is *no* difference between the parent populations of the two samples being compared. Suppose, however, that we set up a consistent alternative to the null hypothesis in the following way. Assume that a small difference *does* exist between the true means of the parent populations, $(\mu_2 - \mu_1)$. If the alternative hypothesis were true, and the observed difference of the means could have occurred with exactly the same probability as could occur if the null hypothesis were true, then we would have no basis for rejecting either hypothesis. Either hypothesis would be equally consistent with the available data. If the difference in means implied by the alternative hypothesis is of scientific significance, it is clear that we have insufficient data to reach a definite conclusion; the experiment is incomplete, and we need to obtain additional measurements.

For example, suppose we take a random sample of five observations from each of the two groups comprising the data of Dr. Kleiber, presented in Chapter 6. In this previous example we used all of the data, and it was possible to reject the null hypothesis. Therefore, we know ahead of time that a real difference probably exists between the two parent populations. However, the analysis summarized in Table 15-1 using the smaller samples fails to reveal a difference that

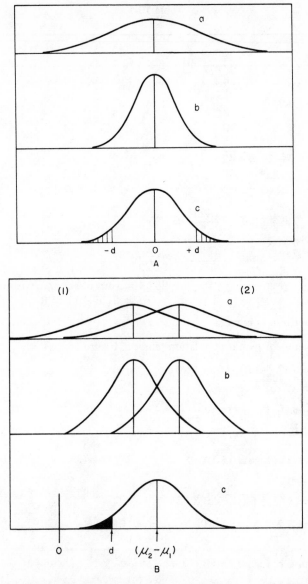

Figure 15-1. (A) Distribution implied by null hypothesis. *a.* Parent population. *b.* Distribution of sample means. *c.* Distribution of differences in means between randomly selected pairs of samples. (B) Distributions implied by consistent alternative hypothesis. *a.* (1) Parent population with mean μ_1· (2) Parent population with mean μ_2. *b.* Distributions of sample means from the parent populations. *c.* Distribution of differences between sample pairs.

Table 15-1

	x_1	x_2
	4	7
	8	10
	8	11
	10	14
	12	26
$\Sigma x =$	42	68
$\Sigma x^2 =$	388	1142
$s =$	2.97	7.37

$\bar{x}_1 = 8.4$

$\bar{x}_2 = 13.6$

$d = 5.2$

$$388 - \frac{(42)^2}{5} = 35.2$$

$$1142 - \frac{(68)^2}{5} = 217.2$$

$\overline{252.4} = $ sum of squares within columns

$\text{df} = 2(5 - 1) = 8$

If null hypothesis is correct:

Parameters of parent population (curve a, Figure 15-1A)

$\bar{\bar{x}} = (42 + 68)/10 = 11.0$

$s' = \sqrt{252.4/8} = 5.62$

For curve b, Figure 15-1A:

$s'/\sqrt{n} = 5.62/\sqrt{5} = 2.51$

For curve c, Figure 15-1A:

$\mu = 0$

$s'' = \sqrt{2(5.62)^2/5} = 3.55$

$t = 5.2/3.55 = 1.46 \qquad (p = \sim 18\%)$

$\qquad\qquad\qquad\qquad (\text{df} = 8)$

For curve c, Figure 15-1B:

$s'' = 3.55$

$$t = \frac{(\mu_2 - \mu_1) - 5.2}{3.55} = 1.46$$

$(\mu_2 - \mu_1) = 10.4$

is not consistent with the null hypothesis. If this were all the data available and we did not know the contents of the remainder of this chapter, we might make the foolish mistake of concluding that there was no difference of any significance between these two populations. It would be small consolation to learn that we were not the first to make this kind of mistake.

From the calculations summarized in Table 15-1, we see that the calculated value of t is 1.46. For 8 degrees of freedom (i.e., $n - 2$), we see from the tables of t distribution that this observed difference could occur by chance somewhere between 10 and 25% of the time. Using a rough graphic interpolation, we see that this could occur by chance around 18% of the time if there were no difference between the two groups.

In other words, the null hypothesis implies the following. There is a single population which we have randomly divided into groups of five. At random we select pairs of groups and record the differences in group means. About 18% of the time the differences observed would be as large or larger than the difference observed between the two groups shown in Table 15-1. The distribution curves in Figure 15-1A illustrate these implications of the null hypothesis. Curve (a) represents a single parent population. Curve (b) shows the distribution of means of groups of five. For curve (c) we assume that in generating a population of differences between pairs of means, we always subtract the mean of the first group selected from the mean of the group making the second member of the pair. Thus, the observed difference in means will be positive just as often as negative. The experimentally observed difference, shown as $\pm d$, can be exceeded by the proportion of the population represented by the shaded tails.

We will now consider an alternative hypothesis by asking how far apart the means of the two populations might be if the probability of our obtaining the observed difference by chance is precisely the same as it would be if there were no difference. In other words, assuming that a difference exists and that the mean of the control population is estimated at μ_1 and the mean of the experimental group at μ_2, is there a value for the difference, $\mu_2 - \mu_1$, greater than zero such that the observed difference could have occurred by chance with the same probability as calculated on the basis of the null hypothesis?

Figure 15-1B illustrates the implications of the alternative hypothesis. Two parent populations with means, μ_1 and μ_2, are shown at

(a) in the figure as curves (1) and (2). After the parent populations are divided into groups of five, their means will be distributed as shown by the curves at (b). We now assume that pairs are formed by selecting one group from each population at random, and the differences between the means in each pair are recorded. In the consideration of null hypothesis, we always subtracted the mean of the first group selected from the mean of the second group. In this instance we always subtract the mean of the group from population (1) from the mean of population (2). The observed differences generate a distribution as shown at (c).

On curve (c) the location of d, the experimentally observed difference, is marked.

Table 15-1 summarizes the calculations involved in fitting the data to the theoretical curves illustrated in Figure 15-1A and B. Note how the variance of the population of differences is the same for both hypotheses. Thus, the two curves have the same shape and differ only in the location of their means. If for a moment we imagine that both populations existed, they could be portrayed side by side as shown in Figure 15-2. The significance of this combined

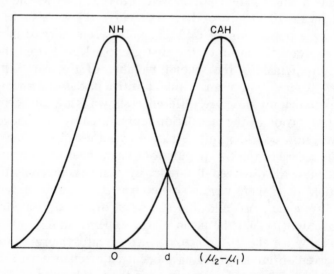

Figure 15-2. Relationship of observed difference to the distribution (NH) implied by the null hypothesis and to the equally consistent alternative hypothesis (CAH).

picture lies in the location of d cutting off equal tails from each of the populations. That is, the observed difference is equally consistent with the postulated population for the null hypothesis (shown as curve NH) or with the postulated population implied by the consistent alternative hypothesis (shown as curve CAH).

Since the observed difference is means, 5.2 days, lies equidistant between zero days of the null hypothesis and the mean of the consistent alternative hypothesis, the latter mean must be equal to 10.4 days. In short, the observed data are equally consistent with a difference of zero days and a difference of 10.4 days. Now 10 or more days difference in this experiment is certainly a biologically significant difference. Also the data which have been made available clearly do not give a basis for selecting one hypothesis over the other. However, it would hardly make sense to suggest that we do not know a little more about the possibilities from the data than we did before beginning the experiment. Our task at this point is to see what can be inferred from the data, meager as they may be.

Using the observed difference is means, 5.2, we can estimate the confidence limits in the usual manner as:

$$\mu = (\text{diff}) \pm ts$$

For this purpose we need to consider the distribution of differences generated by repeated samplings, n_1 and n_2, from each population. As shown in Chapter 13, the variance of a difference is equal to the sum of variances of the two components. Thus, where V_d is the variance of the differences, and V_1 and V_2 are variances from samples of n_1 and n_2, respectively,

$$V_d = V_1/n_1 + V_2/n_2$$

In this example,

$$V_d = \frac{(2.97)^2}{5} + \frac{(7.37)^2}{5} = 12.62$$
$$s_d = 3.55$$

The degrees of freedom are $(n_1 + n_2 - 2)$. Therefore, the appropriate value for t from Table A-3 is 2.31. The confidence limits will be:

$$(\mu_1 - \mu_2) = 5.2 \pm 3.55(2.31)$$
$$= -3.2, + 13.3$$

Therefore, although our data are equally consistent with a difference of zero and 10.4, the 95% confidence limits indicate that the universe mean of these differences may be as low as -3.2 but could also be as high as 13.3.

This upper limit helps determine whether or not the investigation should be continued. If this upper limit is of no practical importance, we may conclude the investigation. We can say with 95% confidence that the difference is less than 13.3 in this instance.

On the other hand, if the minimum level of interest is below the upper confidence limit, continuation of the investigation often requires some estimate of the increased numbers of observations that will be needed. In this example, assume that 5.2 days is chosen as the minimum level of interest.

There are various ways to approach this problem, but in view of the combined uncertainties of the parameters, mean and standard deviation, a crude but simple approximation of some minimum number usually suffices. To repeat the experiment with less than this minimum almost guarantees that the data will still be inconclusive.

In this approximation we assume that the means and variances are not likely to undergo large changes. Some changes with more data are to be expected, but the effects of these changes which appear in the numerators are likely to be small compared to the effect of changes in the n values in the denominators.

The objective of this approximation is to shrink the terms, V_1/n_1 and V_2/n_2, so that the mean difference is at least two standard deviations away from zero. Thus, the new standard deviation here must be equal to half the minimum, 5.2. Then,

$$s = 2.6$$
$$s^2 = 6.76 = \frac{(2.97)^2}{n_1} + \frac{(7.37)^2}{n_2}$$

If n_1 and n_2 are equal, solving for n will indicate here the need for a total of 9 pairs or 18. Notice, however, that one of the variances is much larger than the other. The contribution of each of the component variances is proportional to its size. Therefore, increasing n in the group with the larger variance will be more effective than increasing n qually in both groups. Therefore, we solve for a new n_2 as follows:

$$6.76 = \frac{(2.97)^2}{5} + \frac{(7.37)^2}{n_2}$$
$$n_2 = 11$$

This indicates that increasing the number in the second group to 11 (i.e., adding 6 more) will be as effective as increasing both groups to 9 each (i.e., a total of 8 more).

SUMMARY

Experimental data in which two means are compared may be consistent with the null hypothesis, especially if the data are obtained from a small pilot study. As a general rule, when the difference between the means of the groups is consistent with the null hypothesis, it is equally consistent with a value twice as large as the mean difference obtained in the experiment. With this value and the maximum consistent difference calculated with the appropriate t-test value, the potential scientific or practical value of further investigation can be evaluated.

PROBLEMS

1. An abstract of a journal report states that the mean survival time of leucemic patients treated with agent A was 4 months longer than the mean survival time of those treated with agent B, but that the difference was not statistically significant. What is the consistent alternative hypothesis regarding the possible increase in survival time?
2. Two groups, each containing 10 weanling rats, are fed diets A and B. At the end of 4 months the mean weight of rats on A was 150 grams. The *SD* of this group was 12.5 grams. At the same time the mean weight in group B was 160 grams with a *SD* of 13 grams. The observed difference could have occurred by chance ore than 5% of the time if there were no difference in the diets.
    a. What is the consistent alternative hypothesis?
    b. What is the variance of the differences?
    c. Estimate the number of pairs needed to establish a true difference of at least 10 grams.

16
DISTRIBUTION-FREE METHODS
(NONPARAMETRIC STATISTICS)

With a few exceptions all of the methods presented in previous chapters were based on certain assumptions either explicitly stated or implied. First, it was generally implied that all of the measurements considered could be classified as either continuous or discontinuous variables. There are instances, however, in which measurements do not fall clearly into either of these groups. In addition, there are times when the scales from which apparently continuous variables are derived is nonuniform. This possibility was encountered in Chapter 8 where a nonlinear function was correlated with a linear function by using rank-correlation, a nonparametric method. Some biological measurements are necessarily crude in their early stages of investigation, and it is not rare to find comparisons based on a scale of one to four plus. Thus, for example, urine is tested for protein by adding a few drops of acetic acid to about 5 ml of the urine in a test tube and heating the mixture. If protein is present, a precipitate of coagulated protein forms. The amount of protein present is roughly estimated from the amount of precipitate formed using the following scale: a light cloud formed after heating is graded as one plus; a heavy cloud is graded as two plus, a light coagulum as three plus, and a heavy coagulum as four plus. A measurement of this type does not fit either into the continuous or discontinuous variety of measurements. Furthermore, in many of these instances the difference between 1+ and 2+ may be larger than the difference between 2+ and 3+ but smaller than the difference between 3+ and 4+.

Another assumption, which we have used extensively but which may not be true in many instances, was that the distribution could be readily compared to some well-known form, such as the normal, the binomial, or the Poisson. Fortunately, the distribution of the parameters of samples of nonnormal distributions are sufficiently close to normal to be easily subjected to statistical analysis and appropriate statistical inferences. However, there are numerous occasions when we must deal with inferences based on the parent population's distribution rather than on the distribution of derived parameters. This occurs most commonly when there is insufficient data about a variable to have any idea of its distribution, but there are also occasions when there is sufficient evidence to indicate that the distribution is not normal.

Because of these difficulties there has been a fairly extensive development of nonparametric and distribution-free methods in recent years. In this chapter a few that are most frequently needed in biological and medical investigations will be considered. These methods are largely based on extensions of the binomial distribution and the methods selected for presentation are designed to demonstrate this general approach. Thus if the methods presented here do not clearly apply to a problem encountered by the reader, it is hoped that he will be able to utilize this general approach to find a reasonably reliable method, which adequately deals with the situation. It must be emphasized here, however, that nonparametric methods, being more general than those for which useful assumptions can be made, do not offer the same degree of precision for either estimates or inferences. These methods should therefore be used only when the standard methods are clearly not applicable.

One of the basic procedures for dealing with continuous but nonlinear measurements is the *sign test*. Its essential nature will be revealed by its application in estimating the median of a sample of data. As indicated in Chapter 1 the limitations of the human mind make it necessary for us to find a single number to summarize the general character of a collection of data. For most purposes the average or arithmetic mean is the best parameter. For those collections of data where the mean would be inappropriate or even misleading, the median is usually the best substitute. The median in a set of data is that value which exceeds half the population and is exceeded by the remaining half. In a word, it is the halfway mark when the data are ranked. The halfway mark on the ranked data is the median, m, of the sample, but we may wish to know if our sample is consistent with a parent population whose median is M. To test the hypothesis that the sample could have arisen from the parent population with median, M, we apply the sign test as follows.

If M is the true median of the population, the probability that any given measurement exceeds M is exactly $1/2$, and the probability that M exceeds the measurement is also exactly $1/2$. Thus, if the sample size is n, we determine the probability that as large a proportion of the sample as a/n will be less than the median value. Mathematically, this does not differ from the problem of determining the probability that the proportion of heads will be a/n after n tosses of an unbiased coin.

These principles may be demonstrated with the following problem. Suppose that when we inoculate a large series of culture tubes with a specific strain of bacteria, we find that the time required to decolorize a measured amount of indicator in each of the tubes varies so that the median time required is 72 minutes. Now when a series of ten tubes are inoculated with an unknown but possibly identical strain of bacteria, we find that the times required for each of these tubes to become decolorized are as follows: 59, 60, 61, 63, 64, 67, 70, 75, 83, and 100 minutes, tabulated in ranked order. From this collection of data we see that the median value is exceeded three times in a series of ten. We need to determine the probability that the median will be exceeded as few as three times. Note that this will be a cumulative probability, since it includes the probability that the median will be exceeded zero, one, two, and three times. Using the point binomial, we calculate the following probabilities:

$$p_0 = 1/1024 \qquad p_2 = 45/1024$$
$$p_1 = 10/1024 \qquad p_3 = 120/1024$$

Thus, the cumulative probability is $176/1024$ or a little more than 17%. Since this is considerably more than 5%, we do not have conclusive evidence that the unknown strain is different from the first strain measured. On the other hand, although we have insufficient evidence to rule out the null hypothesis, we do not have very strong evidence that the strains are identical.

To consider this matter further, we need to know what limits of the universe medians this sample could be consistent with. This is easily determined by seeing how far up we can add the individual probabilities, as was done above, without exceeding the 5% level. When we add p_0 and p_1, we have 1.07%, while the addition of one more step, p_2, puts us over the mark at 5.47%. Furthermore, when we attempt to set the fiducial limits of the median, we need a two-sided or double-tailed distribution. Therefore, if we count in two places from both ends of the data, we will be able to state that the probability of the median of the parent population falling outside these limits is less than 2.14% (i.e., $2 \times 1.07\%$). In this case the lower limit is between 60 and 61, while the upper limit lies somewhere between 75 and 83.

These results show that because of the discontinuous character of the binomial distribution, the limits established are apt to be somewhat more stringent than usually required, especially when small samples are being used. With small samples there is, as we have already seen, very little difficulty in calculating the individual probabilities and adding them up as required. With samples only a little larger, the computational labor soon becomes excessive. However, when the sample size reaches 10 or exceeds this amount, the approximations using the normal distribution becomes useful estimates of the limits. When the sample size is between 10 and 15, the error in using this approximation is still too large, but beyond 15 the approximation is close enough for almost all practical purposes.

In previous chapters some form of the t test was generally used in comparing two sample means. We are usually inclined to be particularly confident in the validity of these analyses when the data may be treated as paired samples. If the conditions under which each of the pairs has been observed are identical, our assumption that the distribution of the differences about the mean is equal to the distribution of the parent population about the mean of the population is valid. Occasionally, one pair may be observed under one set of conditions, while another pair is observed under very different conditions. Under these circumstances, it would be highly unlikely that the spread of the distribution would be unchanged. Thus, for example, if a pair of observations or physiological measurements were made on part of a group of frogs in the spring and the remainder of the observations were carried out at various times extending well into the winter season, one would be quite naive to assume that the various pairs of observations could be treated as one population of variables. The implied assumptions of the t test under there circumstances would have little or no validity. Needless to say, the mean difference observed would not be acceptable.

In some biological experiments or observations the intrusion of such additional variation in circumstances is unavoidable. To solve this kind of problem, a form of the sign test can be utilized. The general nature of this test may be described as follows. Let x and y represent pairs of measurements that compare two conditions. The pairs are then: $(x_1, y_1), (x_2, y_2), \cdots, (x_n, y_n)$, when there are n pairs. We can now apply the sign test to the differences generated as: $(x_1 - y_1), (x_2 - y_2), \cdots,$

$(x_n - y_n)$. The null hypothesis here assumes that there is no real difference. If this is so, the number of negative values will be expected to occur the same proportion of the time as the positive values. Note that the problem has been converted to a form of the heads and tails type of binomial distribution. That is, we analyze the proportion of positive and negative results on the basis of the binomial distribution when $p = 1/2$.

The following example demonstrates a variety of difficulties that would prevent the usual application of the t test. In this example the growth of hemolytic streptococci in nutrient media to which measured amounts of serum have been added is graded as 0 to 4+. This experiment is designed to test the adequacy of absorption of an antibiotic in 1 hour. For this purpose, serum samples are obtained before administration of the antibiotic and 1 hour after. Twenty subjects are studied; thus, 20 pairs of culture tubes are available for evaluation. The results are summarized in Table 16-1.

In most applications of this form of analysis, a tie between the members of the pair is usually discarded, and the number of observations is diminished accordingly. In this particular example, a tie reveals the absence of significant inhibitory effect just as effectively as a grossly observable difference. Therefore, wherever ties occur in an experiment of this kind, these are included as negative effects. In this specific problem, we note that grossly observable apparent inhibition occurred in 13 out of 20 samples. If we use the normal approximation for the binomial distribution as a one-sided or single-tailed test, we may proceed as follows:

$$\frac{x - np}{\sqrt{npq}} = \frac{13 - (0.5)(20)}{\sqrt{(20)(0.5)(0.5)}} = \frac{3}{2.24} = 1.34$$

From Table A-2II we find that this difference could occur by chance about 9% of the time and are therefore unable to reject the null hypothesis. That is, the observed differences could be attributed to the chance variations in the behavior of the culture and might be totally unrelated to the antibiotic used.

Usually at this point the same questions that were analyzed in Chapter 15 arise. Specifically, we would like to know how good this antibiotic could possibly be shown to be if we used a much larger series of observations. To answer this question, we assume that the result that we obtained could have been obtained as often as 5% of the time if p represents the proportion of the tubes in which a

Table 16-1

Subject No.	Growth After	Growth Before	Difference
1	4	4	−
2	4	2	−
3	1	2	+
4	3	4	+
5	0	4	+
6	5	2	−
7	1	4	+
8	4	1	−
9	1	3	+
10	1	3	+
11	0	1	+
12	3	4	+
13	2	2	−
14	2	3	+
15	2	2	−
16	3	2	−
17	0	4	+
18	2	3	+
19	1	4	+
20	1	3	+

grossly positive effect could be observed. This may be calculated from the following relationship:

$$1.645 = \frac{np - a}{\sqrt{npq}} = \frac{20p - 13}{\sqrt{20p(1 - p)}}$$

which becomes the quadratic relationship:

$$454p^2 - 574p + 169 = 0$$

from which we obtain $p = 0.8$.

A useful variant of this comparison of paired values is occasionally applicable when a probable significant difference has been demonstrated. Thus, for example, 20 pairs of measurements might be compared and a positive difference found in every instance. By referring to Table A-6 in the appendix, we find that this would be consistent at the 95% confidence level with true values of p ranging from about 0.86 to 1.00. Stated in a general way, we obtained, by comparing the pairs, a series of differences: $(x_1 - y_1)$, $(x_2 - y_2)$, $\cdots$, $(x_n - y_n)$. We now are able to approximate the magnitude of the average difference, D, by finding how large a quantity must be added to the y values in the above expression so that half of the resulting differences become negative in the following series: $(x_1 - y_1 - D)$, $(x_2 - y_2 - D)$, etc. The value for D is easily foung by arranging the observed differences in ranked order; D is the median value. Then, as previously demonstrated, the 95% confidence limits of this median can be found either by binomial summations or by the normal approximation.

In this chapter a few examples of the nonparametric methods, which can be used in biological investigations have been presented. The rank correlation method is another example of a nonparametric method and so also are some of the applications of the χ^2 distribution demonstrated in Chapter 16. This chapter, however, has been limited to those methods which are readily resolved into applications of the binomial. The intent here is to enable the reader to look for a suitable nonparametric solution to problems not amenable to analysis by parametric methods or by the type of methods illustrated here as variants to the sign test.

SUMMARY

Nonparametric methods of analysis are necessary for the evaluation of either semiquantitative, ranked data or quantitative data whose scale may be either nonuniform or unknown. These data are summarized using the median instead of the mean. A given set of data may be compared to the data of the same or some other population by comparing the incidence of values on either side of the median. Paired data may be evaluated by the use of the sign test. In both cases, the data is converted essentially to a problem involving the binomial distribution.

PROBLEMS

1. Five students in a physiology laboratory section perform an experiment in which the blood pressure is measured before and after one hand is immersed in ice water. Their results are tabulated as follows:

Before	After
103	104
110	115
120	127
115	124
120	145

One student using the t-test states that no rise in blood pressure is demonstrated since the mean difference could have been observed by chance more than 25% of the time. A second student states that the experiment demonstrates a rise in blood pressure since the chance probability of all five measurements showing a rise is 1/32. Discuss the arguments for and against each answer. How would you improve the directions in the laboratory manual so that a group of five students would be likely to obtain a definite answer using only each other as experimental subjects?

2. Ten pairs of rats were used to compare two diets. Their weights in grams are given in the following table:

Diet A	Diet B
122	110
120	108
112	94
126	138
112	102
126	122
118	114
112	108
88	124
122	116

Compare the estimates of the probability of obtaining the observed differences by chance using the following methods:
a. Treat as paired data and use the t-test.
b. Treat as unpaired data and use the t-test.
c. Treat as paired data and use the sign test.

EPILOGUE

The purpose of these added paragraphs is to bring into perspective a few notions about probability for which we could find no appropriate chapter. Most of this book emphasizes methods for estimating the probability that some event or series of incidents has occurred by chance. Ordinarily when this probability is less than 5%, random chance is rejected as a possible explanation.

This raises two questions: (1) Does this mean that we are likely to be wrong 5% of the time, and (2) is this critical value of 5% always the best criterion?

The answers are neither yes nor no. It depends. If 100 comparisons are made among which there are no real differences, in *these* comparisons we might be wrong 5% of the time. Most comparisons, however, arise from well-planned experiments in which there are good reasons to expect real differences. In these we test the possibility that the fundamental premise is in error and that the differences observed could be attributed to chance. The 5% criterion tests the adequacy of the data to support the basic premise. Under these conditions the probability of error must be very small.

On the other hand, screening tests make hundreds of comparisons expecting no more than a few positive results. Added to the correct positive results will be 5% of the hundreds that are negative.

Multiple screening tests for the early detection of disease multiply this problem. Because of the establishment of normal boundary values at ±2 standard deviations from the mean, any normal subject's test value has a 0.95 probability of falling within the normal range. The probability of testing as normal on n tests is 0.95^n.

Routine screening panels consist of more than 20 tests. Theoretically, only about 36% (i.e., 0.95^{20}) of normal subjects will be regarded as completely normal, and 64% will be suspected incorrectly of being unwell. This would suggest that we should be having an epidemic of overdiagnosed, normal patients wasting the services of our clinical facilities. Fortunately, this has not happened.

There are two explanations for this better-than-expected result. First, and probably most important, diagnoses are rarely based on a single abnormality, especially when the deviation from normal is small. Second, there is the assumption that distributions that appear to conform closely to a normal distribution near the mean conform equally well remote from the mean. For example, the binomial distribution, $(1/2 + 1/2)^{10}$, appears to conform remarkably well to a normal distribution, but the relative frequency is zero at 3 standard deviations from the mean (i.e., there is no way to toss 10 coins and find 11 heads). Thus, even though 5% of normal individuals may show test results deviating a little more than 2 SD from the mean, virtually no normals will test much beyond. They may show a "borderline abnormality" that must be considered but will be ignored in the absence of additional consistent evidence.

Nevertheless, suppose that a test distribution does indeed follow the normal distribution well beyond 3 SDs from the mean. Regardless of whether the test is used alone or as part of a large panel of tests, if a few thousand subjects are tested within a relatively short time, there will be a substantial number of potentially misleading results. Unless confirmatory diagnostic information can be added, errors will be inevitable. However, very few diagnostic processes are 100% perfect. We continue to use them because they are more often right than wrong and because the benefits outweigh the consequences of error.

The critical level for rejection or acceptance of the null hypothesis at the 5% level is useful in the vast majority of instances, but this need not be regarded as absolute or inviolable. For example, if a given result could have occurred 10% of the time by chance, this does not prove that it did occur by chance. This means only that the evidence can not be accepted without reservation. Whether or not one chooses to accept the evidence would depend on one's judgement of the circumstances. Certainly, if it were possible to gather more data, it would be prudent to do so. There are, however, unusual occasions when this is not possible, and a calculated risk may be unavoidable.

On the other hand, there are occasions when the probability of an occurrence by chance is less than 5%, and the evidence must still be rejected. These are circumstances when we are faced with deciding whether a given result occurred either by chance or because of some alternative cause. If the alternative is judged to be even more unlikely than a chance occurrence, acceptance of the alternative would be irrational.

Sometimes, however, there is more than one alternative to the null hypothesis. Probability calculations are based on the assumption that uncontrollable variables are acting at random. The effect of bias may be subtle and difficult to detect but is a possibility not to be ignored.

Data from tests of psychic powers or extrasensory perception have been offered with calculations of chance probability well below the 5% level. No doubt the results are not due entirely to chance, but the alternative explanations are still not accepted. This only shows that statistical analysis can not be used to prove the unbelievable.

ANSWERS

Chapter 2
1. 8
2. 4
3. When three different kinds of coins are used, the different combinations can be distinguished.
4. 1, 8, 28, 56, 28, 8, 1
5. 0.109 and 0.234

Chapter 3
1. Sequence of random numbers from Table A-1 are shown in parentheses and followed by number selected:

    (10) 10.42   (33) 9.85   (35) 11.18
    (09) 10.63   (76) 9.69   (86) 8.87
    (73) 9.10   (52) 9.51   (36) 7.47
    (25) 8.81   (01) 10.46
2. Median = 9.69
   Mode = 9.64

Chapter 4
1. mean = 10.6; standard deviation = 0.65
2. 9.34 to 11.88 and 8.94 to 12.28
3. 0.27
4. See derivation in chapter on analysis of variance.
5. 30.6%

Chapter 5
1. 95% limits: 10.41 5o 10.82
   99% limits: 9.58 to 11.64
2. mean = 10.03; standard deviation = 1.24,
   95% confidence limits: 9.45 to 10.61
3. 25%

Chapter 6
1. $t = 1.71$, therefore p is greater than 0.05. The superiority of one of the diets is not established.
2. $t = 1.12$; therefore the data do not support the hypothesis.
3. Compare readings of the first four with the last six taken before stimulation. Comparing means yields $t = 3.86$. Therefore the probability of a difference in group means occurring by chance is less than 0.005.

Chapter 7
1. a. $y = 3x$
   b. $y = 3x + 4$
   c. $y = 50 - x$
2. a. Approx. 37.5%
   b. $1/y = 0.00375 - 0.$
3. a. (conc.) $= 60.8$ (Op. dens.) $- 8.49$
   b. SEE $= 0.383$
   c. ± 1.65 mg %

Chapter 8
1. 0.690
2. 0.543
   $y = 0.472x + 60.5$

Chapter 9
1. 0.01, 0.18
2. Between 9 and 24%
3. Less than 1% but no higher than 58%
4. 0.0547

Chapter 10
1. 1.1 to 2.9
2. 0.238
3. 0.227
4. 0.14

Chapter 11
1. a. 28.7 and 6
   b. 14.6 and 7.6
   c. 13.5 and 8.0
2. $X_c^2 = 10.29, X^2 = 11.65$
   (Note: X_c^2 with subscript, c, is the conventional way of indicating that Yate's correction is present.)
   The data are consistent with the idea that rigid enforcement of sanitation is effective.
3. The data do not support the suggestion.
   $X_c^2 = 0.322, X^2 = 0.707$
4. Since $F_{4,7} = (0.01/0.0064) = 1.56$, probability is greater than 10%.

Chapter 12
1. $p = 0.231$

Chapter 13

1.

Source	Sum of squares		d.f.	Var.
Within		0.1625	9	0.01806
Between	(0.001667)4 =	0.0067	2	0.00333
Total		0.1692	11	

$$F_{2,9} = \frac{0.003}{0.181} \quad P \; 0.10$$

2.

Within		0.1625	9	0.01806
Between	(2.0017)4 =	8.0067	2	4.0033
Total		8.1692	11	

$$F_{2,8} = 400 \quad P \; 0.001$$

Chapter 14

1. Variance within days $\Sigma d^2 / n$ = 0.83/22 = 0.038

   Variance between = Variance of daily means/2 = 0.06

Chapter 15

1. 8 months
2. a. Diff. = 20 g
   b. V = 32.53
   c. 65 pairs

Chapter 16

1. Consider a consistent alternative hypothesis. Increase the number of replications on each student instead of relying on a single reading of a difference.
2. a. $p = 0.05$
   b. $p = 0.05$
   c. $p = 0.0546$

APPENDIX

Table A-1 Random Numbers[a]

```
10 09 73 25 33   76 52 01 35 86   34 67 35 48 76   80 95 90 91 17   39 29 27 49 45
37 54 20 48 05   64 89 47 42 96   24 80 52 40 37   20 63 61 04 02   00 82 29 16 65
08 42 26 89 53   19 64 50 93 03   23 20 90 25 60   15 95 33 47 64   35 08 03 36 06
99 01 90 25 29   09 37 67 07 15   38 31 13 11 65   88 67 67 43 97   04 43 62 76 59
12 80 79 99 70   80 15 73 61 47   64 03 23 66 53   98 95 11 68 77   12 17 17 68 33

66 06 57 47 17   34 07 27 68 50   36 69 73 61 70   65 81 33 98 85   11 19 92 91 70
31 06 01 08 05   45 57 18 24 06   35 30 34 26 14   86 79 90 74 39   23 40 30 97 32
85 26 97 76 02   02 05 16 56 92   68 66 57 48 18   73 05 38 52 47   18 62 38 85 79
63 57 33 21 35   05 32 54 70 48   90 55 35 75 48   28 46 82 87 09   83 49 12 56 24
73 79 64 57 53   03 52 96 47 78   35 80 83 42 82   60 93 52 03 44   35 27 38 84 35

98 52 01 77 67   14 90 56 86 07   22 10 94 05 58   60 97 09 34 33   50 50 07 39 98
11 80 50 54 31   39 80 82 77 32   50 72 56 82 48   29 40 52 42 01   52 77 56 78 51
83 45 29 96 34   06 28 89 80 83   13 74 67 00 78   18 47 54 06 10   68 71 17 78 17
88 68 54 02 00   86 50 75 84 01   36 76 66 79 51   90 36 47 64 93   29 60 91 10 62
99 59 46 73 48   87 51 76 49 69   91 82 60 89 28   93 78 56 13 68   23 47 83 41 13

65 48 11 76 74   17 46 85 09 50   58 04 77 69 74   73 03 95 71 86   40 21 81 65 44
80 12 43 56 35   17 72 70 80 15   45 31 82 23 74   21 11 57 82 53   14 38 55 37 63
74 35 09 98 17   77 40 27 72 14   43 23 60 02 10   45 52 16 42 37   96 28 60 26 55
69 91 62 68 03   66 25 22 91 48   36 93 68 72 03   76 62 11 39 90   94 40 05 64 18
09 89 32 05 05   14 22 56 85 14   46 42 75 67 88   96 29 77 88 22   54 38 21 45 98

91 49 91 45 23   68 47 92 76 86   46 16 28 35 54   94 75 08 99 23   37 08 92 00 48
80 33 69 45 98   26 94 03 68 58   70 29 73 41 35   53 14 03 33 40   42 05 08 23 41
44 10 48 19 49   85 15 74 79 54   32 97 92 65 75   57 60 04 08 81   22 22 20 64 13
12 55 07 37 42   11 10 00 20 40   12 86 07 46 97   96 64 48 94 39   28 70 72 58 15
63 60 64 93 29   16 50 53 44 84   40 21 95 25 63   43 65 17 70 82   07 20 73 17 90

61 19 69 04 46   26 45 74 77 74   51 92 43 37 29   65 39 45 95 93   42 58 26 05 27
15 47 44 52 66   95 27 07 99 53   59 36 78 38 48   82 39 61 01 18   33 21 15 94 66
94 55 72 85 73   67 89 75 43 87   54 62 24 44 31   91 19 04 25 92   92 92 74 59 73
42 48 11 62 13   97 34 40 87 21   16 86 84 87 67   03 07 11 20 59   25 70 14 66 70
23 52 37 83 17   73 20 88 98 37   68 93 59 14 16   26 25 22 96 63   05 52 28 25 62

04 49 35 24 94   75 24 63 38 24   45 86 25 10 25   61 96 27 93 35   65 33 71 24 72
00 54 99 76 54   64 05 18 81 59   96 11 96 38 96   54 69 28 23 91   23 28 72 95 29
35 96 31 53 07   26 89 80 93 54   33 35 13 54 62   77 97 45 00 24   90 10 33 93 33
59 80 80 83 91   45 42 72 68 42   83 60 94 97 00   13 02 12 48 92   78 56 52 01 06
46 05 88 52 36   01 39 09 22 86   77 28 14 40 77   93 91 08 36 47   70 61 74 29 41

32 17 90 05 97   87 37 92 52 41   05 56 70 70 07   86 74 31 71 57   85 39 41 18 38
69 23 46 14 06   20 11 74 52 04   15 95 66 00 00   18 74 39 24 23   97 11 89 63 38
19 56 54 14 30   01 75 87 53 79   40 41 92 15 85   66 67 43 68 06   84 96 28 52 07
45 15 51 49 38   19 47 60 72 46   43 66 79 45 43   59 04 79 00 33   20 82 66 95 41
94 86 43 19 94   36 16 81 08 51   34 88 88 15 53   01 54 03 54 56   05 01 45 11 76

98 08 62 48 26   45 24 02 84 04   44 99 90 88 96   39 09 47 34 07   35 44 13 18 80
33 18 51 62 32   41 94 15 09 49   89 43 54 85 81   88 69 54 19 94   37 54 87 30 43
80 95 10 04 06   96 38 27 07 74   20 15 12 33 87   25 01 62 52 98   94 62 46 11 71
79 75 24 91 40   71 96 12 82 96   69 86 10 25 91   74 85 22 05 39   00 38 75 95 79
18 63 33 25 37   98 14 50 65 71   31 01 02 46 74   05 45 56 14 27   77 93 89 19 36

74 02 94 39 02   77 55 73 22 70   97 79 01 71 19   52 52 75 80 21   80 81 45 17 48
54 17 84 56 11   80 99 33 71 43   05 33 51 29 69   56 12 71 92 55   36 04 09 03 24
11 66 44 98 83   52 07 98 48 27   59 38 17 15 39   09 97 33 34 40   88 46 12 33 56
48 32 47 79 28   31 24 96 47 10   02 29 53 68 70   32 30 75 75 46   15 02 00 99 94
69 07 49 41 38   87 63 79 19 76   35 58 40 44 01   10 51 82 16 15   01 84 87 69 38
```

[a] From *A Million Random Digits,* a publication of the Rand Corporation, Santa Monica, California.

Table A-2I Normal Distribution (Double-Tail). Proportion of area lying outside the ordinates through $U = \pm(\mu - x)/\sigma$

U	0.00	0.01	0.02	0.03	0.04	0.05	0.06	0.07	0.08	0.09
0.0	1.0000	0.9920	0.9840	0.9761	0.9681	0.9601	0.9522	0.9442	0.9362	0.9283
0.1	0.9203	0.9124	0.9045	0.8966	0.8887	0.8808	0.8729	0.8650	0.8572	0.8493
0.2	0.8415	0.8337	0.8259	0.8181	0.8103	0.8026	0.7949	0.7872	0.7795	0.7718
0.3	0.7642	0.7566	0.7490	0.7414	0.7339	0.7263	0.7188	0.7114	0.7039	0.6965
0.4	0.6892	0.6818	0.6745	0.6672	0.6599	0.6527	0.6455	0.6384	0.6312	0.6241
0.5	0.6171	0.6101	0.6031	0.5961	0.5892	0.5823	0.5755	0.5687	0.5619	0.5552
0.6	0.5485	0.5419	0.5353	0.5287	0.5222	0.5157	0.5093	0.5029	0.4965	0.4902
0.7	0.4839	0.4777	0.4715	0.4654	0.4593	0.4533	0.4473	0.4413	0.4354	0.4295
0.8	0.4237	0.4179	0.4122	0.4065	0.4009	0.3953	0.3898	0.3843	0.3789	0.3735
0.9	0.3681	0.3628	0.3576	0.3524	0.3472	0.3421	0.3371	0.3320	0.3271	0.3222
1.0	0.3173	0.3125	0.3077	0.3030	0.2983	0.2937	0.2891	0.2846	0.2801	0.2757
1.1	0.2713	0.2670	0.2627	0.2585	0.2543	0.2501	0.2460	0.2420	0.2380	0.2340
1.2	0.2301	0.2263	0.2225	0.2187	0.2150	0.2113	0.2077	0.2041	0.2005	0.1971
1.3	0.1936	0.1902	0.1868	0.1835	0.1802	0.1770	0.1738	0.1707	0.1676	0.1645
1.4	0.1615	0.1585	0.1556	0.1527	0.1499	0.1471	0.1443	0.1416	0.1389	0.1362
1.5	0.1336	0.1310	0.1285	0.1260	0.1236	0.1211	0.1188	0.1164	0.1141	0.1118
1.6	0.1096	0.1074	0.1052	0.1031	0.1010	0.0989	0.0969	0.0949	0.0930	0.0910
1.7	0.0891	0.0873	0.0854	0.0836	0.0819	0.0801	0.0784	0.0767	0.0751	0.0735
1.8	0.0719	0.0703	0.0688	0.0672	0.0658	0.0643	0.0629	0.0615	0.0601	0.0588
1.9	0.0574	0.0561	0.0549	0.0536	0.0524	0.0512	0.0500	0.0488	0.0477	0.0466
2.0	0.0455	0.0444	0.0434	0.0424	0.0414	0.0404	0.0394	0.0385	0.0375	0.0366
2.1	0.0357	0.0349	0.0340	0.0332	0.0324	0.0316	0.0308	0.0300	0.0293	0.0285
2.2	0.0278	0.0271	0.0264	0.0257	0.0251	0.0244	0.0238	0.0232	0.0226	0.0220
2.3	0.0214	0.0209	0.0203	0.0198	0.0193	0.0188	0.0183	0.0178	0.0173	0.0168
2.4	0.0164	0.0160	0.0155	0.0151	0.0147	0.0143	0.0139	0.0135	0.0131	0.0128
2.5	0.0124	0.0121	0.0117	0.0114	0.0111	0.0108	0.0105	0.0102	0.0099	0.0096
2.6	0.0093	0.0091	0.0088	0.0085	0.0083	0.0081	0.0078	0.0076	0.0074	0.0072
2.7	0.0069	0.0067	0.0065	0.0063	0.0061	0.0060	0.0058	0.0056	0.0054	0.0053
2.8	0.0051	0.0050	0.0048	0.0047	0.0045	0.0044	0.0042	0.0041	0.0040	0.0039
2.9	0.0037	0.0036	0.0035	0.0034	0.0033	0.0032	0.0031	0.0030	0.0029	0.0028
3.0	0.0027	0.0026	0.0025	0.0024	0.0024	0.0023	0.0022	0.0021	0.0021	0.0020

Table A-2II   Normal Distribution (Single-Tail). Proportion of area lying to right of ordinate through $Z = \pm (x - \mu)/\sigma$[a]

Z	0.00	0.01	0.02	0.03	0.04	0.05	0.06	0.07	0.08	0.09
0.0	0.5000	0.4960	0.4920	0.4880	0.4840	0.4801	0.4761	0.4721	0.4681	0.4641
0.1	0.4602	0.4562	0.4522	0.4483	0.4443	0.4404	0.4364	0.4325	0.4286	0.4247
0.2	0.4207	0.4168	0.4129	0.4090	0.4052	0.4013	0.3974	0.3936	0.3897	0.3859
0.3	0.3821	0.3783	0.3745	0.3707	0.3669	0.3632	0.3594	0.3557	0.3520	0.3483
0.4	0.3446	0.3409	0.3372	0.3336	0.3300	0.3264	0.3228	0.3192	$.3156	0.3121
0.5	0.3085	0.3050	0.3015	0.2981	0.2946	0.2912	0.2877	0.2843	0.2810	0.2776
0.6	0.2743	0.2709	0.2676	0.2643	0.2611	0.2578	0.2546	0.2514	0.2483	0.2451
0.7	0.2420	0.2389	0.2358	0.2327	0.2297	0.2266	0.2236	0.2206	0.2177	0.2148
0.8	0.2119	0.2090	0.2061	0.2033	0.2005	0.1977	0.1949	0.1922	0.1894	0.1867
0.9	0.1841	0.1814	0.1788	0.1762	0.1736	0.1711	0.1685	0.1660	0.1635	0.1611
1.0	0.1587	0.1562	0.1539	0.1515	0.1492	0.1469	0.1446	0.1423	0.1401	0.1379
1.1	0.1357	0.1335	0.1314	0.1292	0.1271	0.1251	0.1230	0.1210	0.1190	0.1170
1.2	0.1151	0.1131	0.1112	0.1093	0.1075	0.1056	0.1038	0.1020	0.1003	0.0985
1.3	0.0968	0.0951	0.0934	0.0918	0.0901	0.0885	0.0869	0.0853	0.0838	0.0823
1.4	0.0808	0.0793	0.0778	0.0764	0.0749	0.0735	0.0721	0.0708	0.0694	0.0681
1.5	0.0668	0.0655	0.0643	0.0630	0.0618	0.0606	0.0594	0.0582	0.0571	0.0559
1.6	0.0548	0.0537	0.0526	0.0516	0.0505	0.0495	0.0485	0.0475	0.0465	0.0455
1.7	0.0446	0.0436	0.0427	0.0418	0.0409	0.0401	0.0392	0.0384	0.0375	0.0367
1.8	0.0359	0.0351	0.0344	0.0336	0.0329	0.0322	0.0314	0.0307	0.0301	0.0294
1.9	0.0287	0.0281	0.0274	0.0268	0.0262	0.0256	0.0250	0.0244	0.0239	0.0233
2.0	0.0228	0.0222	0.0217	0.0212	0.0207	0.0202	0.0197	0.0192	0.0188	0.0183
2.1	0.0179	0.0174	0.0170	0.0166	0.0162	0.0158	0.0154	0.0150	0.0146	0.0143
2.2	0.0139	0.0136	0.0132	0.0129	0.0125	0.0122	0.0119	0.0116	0.0113	0.0110
2.3	0.0107	0.0104	0.0102	0.0099	0.0096	0.0094	0.0091	0.0089	0.0087	0.0084
2.4	0.0082	0.0080	0.0078	0.0075	0.0073	0.0071	0.0069	0.0068	0.0066	0.0064
2.5	0.0062	0.0060	0.0059	0.0057	0.0055	0.0054	0.0052	0.0051	0.0049	0.0048
2.6	0.0047	0.0045	0.0044	0.0043	0.0041	0.0040	0.0039	0.0038	0.0037	0.0036
2.7	0.0035	0.0034	0.0033	0.0032	0.0031	0.0030	0.0029	0.0028	0.0027	0.0026
2.8	0.0026	0.0025	0.0024	0.0023	0.0023	0.0022	0.0021	0.0021	0.0020	0.0019
2.9	0.0019	0.0018	0.0018	0.0017	0.0016	0.0016	0.0015	0.0015	0.0014	0.0014
3.0	0013	0.0013	0.0013	0.0012	0.0012	0.0012	0.0011	0.0011	0.0010	0.0010

[a] Tables A-2I and A-2II are taken from Fisher and Yates *Statistical Tables for Biological, Agricultural and Medical Research* published by Oliver & Boyd Ltd., Edinburgh, and by permission of the authors and publishers.

Table A-3 Percentage Points of the t-distribution. ϕ is the number of degrees of freedom.[a]

ϕ	P(%)								ϕ
	50	25	10	5	2.5	1	0.5	0.1	
1	1.00	2.41	6.31	12.7	25.5	63.7	127	637	1
2	0.816	1.60	2.92	4.30	6.21	9.92	14.1	31.6	2
3	0.765	1.42	2.35	3.18	4.18	5.84	7.45	12.9	3
4	0.741	1.34	2.13	2.78	3.50	4.60	5.60	8.61	4
5	0.727	1.30	2.01	2.57	3.16	4.03	4.77	6.86	5
6	0.718	1.27	1.94	2.45	2.97	3.71	4.32	5.96	6
7	0.711	1.25	1.89	2.36	2.84	3.50	4.03	5.40	7
8	0.706	1.24	1.86	2.31	2.75	3.36	3.83	5.04	8
9	0.703	1.23	1.83	2.26	2.68	3.25	3.69	4.78	9
10	0.700	1.22	1.81	2.23	2.63	3.17	3.58	4.59	10
11	0.697	1.21	1.80	2.20	2.59	3.11	3.50	4.44	11
12	0.695	1.21	1.78	2.18	2.56	3.05	3.43	4.32	12
13	0.694	1.20	1.77	2.16	2.53	3.01	3.37	4.22	13
14	0.692	1.20	1.76	2.14	2.51	2.98	3.33	4.14	14
15	0.691	1.20	1.75	2.13	2.49	2.95	3.29	4.07	15
16	0.690	1.19	1.75	2.12	2.47	2.92	3.25	4.01	16
17	0.689	1.19	1.74	2.11	2.46	2.90	3.22	3.96	17
18	0.688	1.19	1.73	2.10	2.44	2.88	3.20	3.92	18
19	0.688	1.19	1.73	2.09	2.43	2.86	3.17	3.88	19
20	0.687	1.18	1.72	2.09	2.42	2.85	3.15	3.85	20
21	0.686	1.18	1.72	2.08	2.41	2.83	3.14	3.82	21
22	0.686	1.18	1.72	2.07	2.41	2.82	3.12	3.79	22
23	0.685	1.18	1.71	2.07	2.40	2.81	3.10	3.77	23
24	0.685	1.18	1.71	2.06	2.39	2.80	3.09	3.74	24
25	0.684	1.18	1.71	2.06	2.38	2.79	3.08	3.72	25
26	0.684	1.18	1.71	2.06	2.38	2.78	3.07	3.71	26
27	0.684	1.18	1.70	2.05	2.37	2.77	3.06	3.69	27
28	0.683	1.17	1.70	2.05	2.37	2.76	3.05	3.67	28
29	.0683	1.17	1.70	2.05	2.36	2.76	3.04	3.66	29
30	0.683	1.17	1.70	2.04	2.36	2.75	3.03	3.65	30
40	0.681	1.17	1.68	2.02	2.33	2.70	2.97	3.55	40
60	0.679	1.16	1.67	2.00	2.30	2.66	2.91	3.46	60
120	0.677	1.16	1.66	1.98	2.27	2.62	2.86	3.37	120
∞	0.674	1.15	1.64	1.96	2.24	2.58	2.81	3.29	∞

[a] Table A-3 is taken from Fisher and Yates *Statistical Tables for Biological, Agricultural and Medical Research* published by Oliver & Boyd Ltd., Edinburgh, and by permission of the authors and publishers.

**Table A-4 Values of Correlation Coefficient for Different Levels of Significance.
ϕ is the number of degrees of freedom.**[a]

ϕ	$P(\%)$					ϕ	$P(\%)$				
	10	5	2	1	0.1		10	5	2	1	0.1
1	0.988	0.997	1.00	1.00	1.00	16	0.400	0.468	0.542	0.590	0.708
2	0.900	0.950	0.980	0.990	0.999	17	0.389	0.455	0.528	5.575	0.693
3	0.805	0.878	0.934	0.959	0.991	18	0.378	0.444	0.515	0.561	0.679
4	0.729	0.811	0.882	0.917	0.974	19	0.369	0.433	0.503	0.549	0.665
5	0.669	0.754	0.833	0.874	0.951	20	0.360	0.423	0.492	0.537	0.652
6	0.621	0.707	0.789	0.834	0.925	25	0.323	0.381	0.445	0.487	0.597
7	0.582	0.666	0.750	0.798	0.898	30	0.296	0.349	0.409	0.449	0.554
8	0.549	0.632	0.715	0.765	0.872	35	0.275	0.325	0.381	0.418	0.519
9	0.521	0.602	0.685	0.735	0.847	40	0.257	0.304	0.358	0.393	0.490
10	0.497	0.576	0.658	0.708	0.823	45	0.243	0.287	0.338	0.372	0.465
11	0.476	0.553	0.634	0.683	0.801	50	0.231	0.273	0.322	0.354	0.443
12	0.457	0.532	0.612	0.661	0.780	60	0.211	0.250	0.295	0.325	0.408
13	0.441	0.514	0.592	0.641	0.760	70	0.195	0.232	0.274	0.302	0.380
14	0.426	0.497	0.574	0.623	0.742	80	0.183	0.217	0.256	0.283	0.357
15	0.412	0.482	0.558	0.605	0.725	90	0.173	0.205	0.242	0.267	0.337
						100	0.164	0.195	0.230	0.254	0.321

[a] Table A-4 is taken from Fisher and Yates *Statistical Tables for Biological, Agricultural and Medical Research* published by Oliver & Boyd Ltd., Edinburgh, and by permission of the authors and publishers.

Table A-5 Rank Correlation Coefficient[a]

n	Significance Level	
	0.05	0.01
4	1.000	
5	0.900	1.000
6	0.829	0.943
7	0.714	0.893
8	0.643	0.833
9	0.600	0.783
10	0.564	0.746
12	0.504	0.701
14	0.456	0.645
16	0.425	0.601
18	0.399	0.564
20	0.377	0.534
22	0.359	0.508
24	0.343	0.485
26	0.329	0.465
28	0.317	0.448
30	0.306	0.432

[a] Reprinted by permission of the editor of *The Annals of Mathematical Statistics* from "Tabulated Values for Rank Correlation" by E. G. Olds in Vol. IX of the Annals for 1938.

Table A-6-I

x\n	1		2		3		4	
0	.00	.975	.00	.84	.00	.71	.00	.60
1	.025	1.00	.01	.99	.01	.91	.01	.81
2			.16	1.00	.09	.99	.07	.99
3					.29	1.00	.19	.99
4							.40	1.00

x\n	5		6		7		8	
0	.00	.52	.00	.46	.00	.41	.00	.37
1	.01	.72	.00+	.64	.00+	.58	.00+	.53
2	.05	.85	.05	.78	.04	.71	.03	.65
3	.15	.95	.12	.88	.10	.82	.09	.76
4	.28	.99	.22	.96	.18	.90	.16	.84
5	.48	1.00	.36	.99+	.29	.96	.24	.91
6			.54	1.00	.42	.99+	.35	.97
7					.59	1.00	.47	.99+
8							.63	1.00

x\n	9		10		11		12	
0	.00	.34	.00	.31	.00	.28	.00	.26
1	.00+	.48	.00+	.45	.00+	.41	.00+	.38
2	.03	.60	.03	.56	.02	.52	.02	.48
3	.07	.70	0.7	.65	.06	.61	.05	.57
4	.14	.79	.12	.74	.11	.69	.10	.65
5	.21	.86	.19	.81	.17	.77	.15	.72
6	.30	.93	.26	.88	.23	.83	.21	.79
7	.40	.97	.13	.93	.31	.89	.28	.85
8	.42	.99+	.44	.97	.39	.94	.35	.90
9	.66	1.00	.56	.99+	.48	.98	.43	.95
10			.69	1.00	.59	.99+	.52	.98
11					.72	1.00	.62	.99+
12							.74	1.00

x\n	13		14		15		16	
0	.00	.25	.00	.23	.00	.22	.00	.21
1	.00+	.36	.00+	.34	.00+	.32	.00+	.30
2	.02	.45	.02	.43	.02	.40	.02	.38
3	.05	.54	.05	.51	.04	.48	.04	.46
4	.09	.61	.08	.58	.08	.55	.07	.52
5	.14	.68	.13	.65	.12	.62	.11	.59
6	.19	.75	.18	.71	.16	.68	.15	.65
7	.25	.81	.23	.77	.21	.71	.20	.70
8	.32	.86	.29	.82	.27	.79	.25	.75
9	.39	.91	.35	.87	.32	.84	.30	.80
10	.46	.95	.42	.92	.38	.88	.35	.85
11	.55	.98	.49	.95	.45	.92	.41	.89
12	.64	.99+	.57	.98	.52	.96	.48	.93
13	.75	1.00	.66	.99+	.60	.98	.54	.97
14			.77	1.00	.68	.99+	.62	.98
15					.78	1.00	.70	.99+
16							.79	1.00

Tables A-6-I and A-6-II — 95% confidence limits for sample ratios. Double tail distribution in A-6-I; single tail distribution in A-6-II. Lower limit shown in left hand column; upper limit shown in right hand column.

Table A-6-I (Continued)

x\\n	17		18		19		20	
0	.00	.20	.00	.19	.00	.18	.00	.17
1	.00+	.29	.00+	.27	.00+	.27	.00+	.25
2	.01	.36	.01	.35	.01	.33	.01	.32
3	.04	.43	.05	.41	.03	.40	.03	.38
4	.07	.50	.06	.48	.06	.46	.06	.44
5	.10	.56	.10	.53	.09	.51	.09	.49
6	.14	.62	.13	.59	.13	.57	.12	.54
7	.18	.67	.17	.64	.16	.62	.14	.59
8	.23	.72	.22	.69	.20	.67	.19	.64
9	.28	.77	.27	.74	.24	.71	.23	.68
10	.33	.82	.31	.78	.29	.76	.27	.73
11	.38	.86	.36	.83	.34	.80	.32	.77
12	.44	.90	.41	.89	.38	.84	.36	.81
13	.50	.93	.47	.90	.43	.87	.41	.85
14	.57	.96	.52	.94	.49	.91	.46	.88
15	.64	.99	.59	.96	.54	.94	.51	.91
16	.71	.99+	.65	.99	.60	.97	.56	.94
17	.80	1.00	.73	.99+	.67	.99	.62	.97
18			.81	1.00	.74	.99+	.68	.99
19					.82	1.00	.75	.99+
20							.83	1.00

x\\n	21		22		23		24	
0	.00	.16	.00	.15	.00	.15	.00	.14
1	.00+	.24	.00+	.23	.00+	.22	.00+	.21
2	.01	.30	.01	.29	.01	.28	.01	.27
3	.03	.36	.03	.35	.03	.34	.03	.32
4	.05	.42	.05	.40	.05	.39	.05	.37
5	.08	.47	.08	.45	.07	.44	.07	.42
6	.11	.52	.11	.50	.10	.48	.10	.47
7	.15	.57	.14	.55	.13	.53	.13	.51
8	.18	.62	.17	.59	.16	.57	.16	.55
9	.22	.66	.21	.64	.20	.61	.19	.59
10	.26	.70	.24	.68	.23	.66	.22	.63
11	.30	.74	.28	.72	.27	.69	.26	.67
12	.34	.78	.32	.76	.31	.73	.29	.71
13	.38	.82	.36	.79	.34	.77	.33	.74
14	.43	.85	.41	.83	.39	.80	.37	.78
15	.48	.89	.45	.86	.43	.84	.41	.81
16	.53	.92	.50	.89	.47	.87	.45	.84
17	.58	.95	.55	.92	.52	.90	.49	.87
18	.64	.97	.60	.95	.56	.93	.53	.90
19	.70	.99	.65	.97	.61	.95	.58	.93
20	.76	.99+	.71	.99	.66	.97	.63	.95
21	.84	1.00	.77	.99+	.72	.99	.68	.97
22			.85	1.00	.78	.99±	.73	.99
23					.85	1.00	.79	.99+
24							.86	1.00

Table A-6-I (Continued)

x \ n	25		26		27		28	
0	.00	.14	.00	.13	.00	.13	.00	.12
1	.00+	.20	.00+	.20	.00+	.19	.00+	.18
2	.01	.26	.01	.24	.01	.24	.01	.24
3	.03	.31	.02	.30	.02	.29	.02	.28
4	.05	.36	.04	.35	0.4	.34	.04	.33
5	.07	.41	.07	.39	.06	.38	.06	.37
6	.09	.45	.09	.44	.09	.42	.08	.41
7	.12	.49	.12	.48	.11	.46	.11	.45
8	.15	.54	.14	.52	.14	.50	.13	.49
9	.18	.57	.17	.56	.17	.54	.16	.52
10	.21	.61	.20	.59	.19	.58	.19	.56
11	.24	.65	.23	.63	.22	.61	.22	.59
12	.28	.69	.27	.67	.25	.65	.24	.63
13	.31	.72	.30	.70	.29	.68	.28	.66
14	.35	.76	.33	.73	.32	.71	.31	.69
15	.39	.79	.37	.77	.35	.74	.34	.72
16	.43	.82	.41	.80	.39	.78	.37	.76
17	.47	.85	.44	.83	.42	.81	.41	.79
18	.51	.88	.48	.86	.46	.83	.44	.81
19	.55	.91	.52	.88	.50	.86	.48	.84
20	.59	.93	.56	.91	.53	.89	.51	.87
21	.64	.95	.61	.93	.58	.91	.55	.89
22	.69	.97	.65	.96	.62	.94	.59	.92
23	.74	.99	.70	.98	.66	.96	.63	.94
24	.80	.99+	.75	.99	.71	.98	.67	.96
25	.86	1.00	.80	.99+	.76	.99	.72	.98
26			.87	1.00	.81	.99+	.77	.99
27					.87	1.00	.82	.99+
28							.88	1.00

Table A-6-I (Continued)

x \ n	29		30	
0	.00	.12	.00	.12
1	.00+	.18	.00+	.17
2	.01	.23	.01	.22
3	.02	.27	.02	.27
4	.04	.32	.04	.41
5	.06	.36	.06	.35
6	.08	.40	.08	.39
7	.10	.44	.10	.42
8	.13	.47	.12	.46
9	.15	.51	.15	.49
10	.18	.54	.17	.53
11	.21	.58	.20	.56
12	.24	.61	.23	.59
13	.26	.64	.25	.63
14	.29	.67	.28	.66
15	.33	.71	.31	.69
16	.36	.74	.34	.72
17	.39	.67	.37	.75
18	.42	.69	.41	.77
19	.46	.82	.44	.80
20	.49	.85	.47	.83
21	.53	.87	.51	.85
22	.56	.90	.54	.88
23	.60	.92	.58	.90
24	.64	.94	.61	.92
25	.68	.96	.65	.94
26	.73	.98	.69	.96
27	.77	.99	.73	.98
28	.82	.99+	.78	.99
29	.88	1.00	.83	.99+
30			.88	1.00

Table A-6-II

n	1		2		3		4	
x								
0	0.00	0.95	0.00	0.78	0.00	0.63	0.00	0.53
1	0.05	1.00	0.03	0.98	0.02	0.87	0.01	0.76
2			0.23	1.00	0.14	0.98	0.10	0.90
3					0.37	1.00	0.24	0.99
4							0.47	1.00

n	5		6		7		8	
x								
0	0.00	0.45	0.00	0.393	0.00	0.35	0.00	0.31
1	0.01	0.66	0.01	0.58	0.01	0.52	0.01	0.47
2	0.08	0.81	0.07	0.73	0.06	0.66	0.05	0.60
3	0.19	0.92	0.15	0.85	0.13	0.78	0.11	0.71
4	0.34	0.99	0.27	0.93	0.23	0.87	0.19	0.81
5	0.55	1.00	0.42	0.99	0.34	0.94	0.29	0.89
6			0.61	1.00	0.48	0.99	0.40	0.95
7					0.65	1.00	0.53	0.99
8							0.69	1.00

n	9		10		11		12	
x								
0	0.00	0.28	0.00	0.26	0.00	0.24	0.00	0.22
1	0.01	0.43	0.01	0.39	0.01	0.36	0.00	0.34
2	0.05	0.55	0.04	0.51	0.04	0.47	0.03	0.44
3	0.10	0.66	0.09	0.61	0.08	0.56	0.07	0.53
4	0.17	0.75	0.15	0.70	0.13	0.65	0.13	0.61
5	0.25	0.83	0.22	0.78	0.20	0.73	0.18	0.69
6	0.34	0.90	0.30	0.85	0.27	0.80	0.25	0.75
7	0.45	0.95	0.39	0.91	0.35	0.87	0.32	0.82
8	0.57	1.00	0.49	0.96	0.44	0.92	0.39	0.88
9	0.72	1.00	0.61	1.00	0.53	0.97	0.47	0.93
10			0.74	1.00	0.64	1.00	0.56	0.97
11					0.76	1.00	0.66	1.00
12							0.78	1.00

n	13		14		15		16	
x								
0	0.00	0.21	0.00	0.19	0.00	0.18	0.00	0.17
1	0.00	0.32	0.00	0.30	0.00	0.28	0.00	0.27
2	0.03	0.41	0.03	0.39	0.03	0.36	0.02	0.35
3	0.07	0.49	0.06	0.47	0.06	0.44	0.06	0.42
4	0.11	0.57	0.10	0.54	0.10	0.51	0.09	0.49
5	0.17	0.65	0.15	0.61	0.14	0.58	0.13	0.55
6	0.22	0.71	0.21	0.67	0.19	0.64	0.18	0.61
7	0.29	0.78	0.26	0.74	0.24	0.70	0.23	0.67
8	0.35	0.84	0.33	0.79	0.30	0.76	0.28	0.72
9	0.43	0.89	0.39	0.85	0.36	0.81	0.33	0.77
10	0.51	0.93	0.46	0.90	0.42	0.86	0.39	0.82
11	0.59	0.97	0.54	0.94	0.49	0.90	0.45	0.87
12	0.68	1.00	0.62	0.97	0.56	0.94	0.52	0.91
13	0.79	1.00	0.70	1.00	0.64	0.97	0.58	0.95
14			0.81	1.00	0.72	1.00	0.66	0.98
15					0.82	1.00	0.74	1.00
16							0.83	1.00

Table A-6-II (Continued)

n	17		18		19		20	
x								
0	0.00	0.16	0.00	0.154	0.00	0.15	0.00	0.14
1	0.00	0.25	0.00	0.24	0.00	0.23	0.00	0.22
2	0.02	0.33	0.02	0.31	0.02	0.30	0.02	0.28
3	0.05	0.40	0.05	0.38	0.05	0.36	0.04	0.34
4	0.08	0.46	0.08	0.44	0.08	0.42	0.07	0.40
5	0 12	0.51	0.12	0.50	0.11	0.48	0.11	0.46
6	0.17	0.58	0.16	0.55	0.15	0.53	0.14	0.51
7	0.21	0.64	0.20	0.61	0.19	0.58	0.18	0.56
8	0.26	0.69	0.24	0.66	0.23	0.63	0.22	0.61
9	0.31	0.74	0.29	0.71	0.26	0.68	0.26	0.65
10	0.36	0.79	0.34	0.76	0.32	0.74	0.31	0.70
11	0.42	0.83	0.39	0.80	0.37	0.77	0.35	0.74
12	0.49	0.88	0.45	0.84	0.42	0.81	0.39	0.78
13	0.54	0.92	0.50	0.88	0.47	0.85	0.44	0.82
14	0.61	0.95	0.56	0.92	0.52	0.89	0.49	0.86
15	0.67	0.98	0.62	0.95	0.58	0.93	0.54	0.89
16	0.75	1.00	0.69	0.98	0.64	0.95	0.60	0.93
17	0.84	1.00	0.76	1.00	0.70	0.98	0.66	0.96
18			0.85	1.00	0.77	1.00	0.72	0.98
19					0.86	1.00	0.78	1.00
20							0.86	1.00

n	21		22		23		24		25	
x										
0	0.00	0.13	0.00	0.13	0.00	0.12	0.00	0.12	0.00	0.11
1	0.00	0.21	0.00	0.20	0.00	0.19	0.00	0.18	0.00	0.18
2	0.02	0.27	0.02	0.26	0.02	0.25	0.01	0.24	0.01	0.23
3	0.04	0.33	0.04	0.32	0.04	0.30	0.04	0.29	0.03	0.28
4	0.07	0.38	0.07	0.37	0.06	0.36	0.06	0.34	0.06	0.33
5	0.10	0.44	0.09	0.42	0.09	0.40	0.09	0.39	0.08	0.38
6	0.13	0.49	0.13	0.47	0.12	0.45	0.11	0.44	0.11	0.42
7	0.17	0.54	0.16	0.52	0.15	0.50	0.15	0.48	0.14	0.46
8	0.21	0.58	0.20	0.56	0.19	0.54	0.18	0.52	0.17	0.50
9	0.25	0.63	0.23	0.61	0.22	0.58	0.21	0.56	0.20	0.54
10	0.29	0.67	0.27	0.65	0.26	0.63	0.25	0.60	0.24	0.58
11	0.33	0.71	0.31	0.69	0.30	0.67	0.28	0.64	0.27	0.62
12	0.37	0.76	0.35	0.73	0.34	0.70	0.31	0.69	0.31	0.66
13	0.42	0.80	0.40	0.77	0.38	0.74	0.36	0.72	0.34	0.70
14	0.46	0.83	0.44	0.80	0.42	0.78	0.40	0.75	0.38	0.73
15	0.51	0.87	0.48	0.84	0.46	0.81	0.44	0.79	0.42	0.76
16	0.56	0.90	0.53	0.87	0.50	0.85	0.48	0.82	0.46	0.80
17	0.62	0.93	0.58	0.91	0.55	0.88	0.52	0.85	0.50	0.83
18	0.67	0.96	0.63	0.94	0.60	0.91	0.57	0.89	0.54	0.86
19	0.73	0.98	0.68	0.96	0.65	0.94	0.61	0.91	0.58	0.89
20	0.79	1.00	0.74	0.98	0.70	0.96	0.66	0.94	0.62	0.92
21	0.87	1.00	0.80	1.00	0.75	0.99	0.71	0.96	0.67	0.94
22			0.87	1.00	0.81	1.00	0.76	0.99	0.72	0.97
23					0.88	1.00	0.82	1.00	0.77	0.99
24							0.88	1.00	0.82	1.00
25									0.89	1.00

Table A-6-II (Continued)

n x	26		27		28		
0	.00	.11	.00	.11	.00	.10	
1	.00+	.11	.00+	.16	.00+	.17	
2	.01	.22	.01	.22	.01	.21	
3	.03	.27	.03	.26	.03	.25	
4	.05	.32	.05	.31	.05	.30	
5	.08	.36	.08	.35	.07	.34	
6	.11	.41	.10	.39	.10	.38	
7	.13	.45	.13	.43	.12	.42	
8	.16	.49	.16	.47	.15	.46	
9	.19	.53	.19	.51	.18	.49	
10	.23	.56	.22	.55	.21	.53	
11	.26	.60	.25	.58	.24	.57	
12	.29	.64	.28	.62	.27	.60	
13	.33	.67	.31	.65	.30	.63	
14	.36	.71	.35	.69	.33	.67	
15	.40	.74	.38	.72	.37	.70	
16	.44	.77	.42	.78	.43	.76	
18	.51	.84	.49	.81	.47	.79	
19	55	.87	.53	.84	.51	.82	
20	.59	,89	.57	.87	.54	.85	
21	.64	.92	.61	.90	.58	.88	
22	.68	.95	.65	.92	.62	.90	
23	.73	.97	.69	.95	.66	.93	
24	.78	.99	.74	.97	.70	.95	
25	.83	.99+	.78	.99	.75	.97	
26	.89	1.00	.84	.99+	.79	.99	
27			.89	1.00	.84	.99+	
28					.90	1.00	

Table A-6-II (Continued)

n x	29		30	
0	.00	.10	.00	.10
1	.00+	.15	.00+	.15
2	.01	.20	.01	.20
3	.03	.25	.03	.24
4	.05	.29	.05	.28
5	.07	.33	.07	.32
6	.09	.37	.09	.36
7	.12	.41	.11	.39
8	.14	.44	.14	.43
9	.17	.48	.17	.47
10	.20	.51	.19	.50
11	.23	.55	.22	.53
12	.26	.58	.25	.57
13	.29	.62	.28	.60
14	.32	.65	.31	.63
15	.35	.68	.34	.66
16	.38	.71	.37	.69
17	.42	.74	.40	.72
18	.45	.77	.43	.75
19	.49	.80	.47	.78
20	.52	.83	.50	.81
21	.56	.86	.53	.83
22	.59	.88	.57	.86
23	.63	.91	.61	.89
24	.67	.93	.64	.91
25	.71	.95	.68	.93
26	.75	.97	.72	.95
27	.80	.99	.76	.97
28	.85	.99+	.80	.99
29	.90	1.00	.85	.99+
30			.90	1.00

Table A-7

ϕ	P(%) 99.5	99	97.5	95	90	75	50	25	10	5	2.5	1	0.5	0.1	ϕ
1	—	—	—	—	0.016	0.102	0.455	1.32	2.71	3.84	5.02	6.63	7.88	10.8	1
2	0.010	0.020	0.051	0.103	0.211	0.575	1.39	2.77	4.61	5.99	7.38	9.21	10.6	13.8	2
3	0.072	0.115	0.216	0.352	0.584	1.21	2.37	4.11	6.25	7.81	9.35	11.3	12.8	16.3	3
4	0.207	0.297	0.484	0.711	1.06	1.92	3.36	5.39	7.78	9.49	11.1	13.3	14.9	18.5	4
5	0.412	0.554	0.831	1.15	1.61	2.67	4.35	6.63	9.24	11.1	12.8	15.1	16.7	20.5	5
6	0.676	0.872	1.24	1.64	2.20	3.45	5.35	7.84	10.6	12.6	14.4	16.8	18.5	22.5	6
7	0.989	1.24	1.69	2.17	2.83	4.25	6.35	9.04	12.0	14.1	16.0	18.5	20.3	24.3	7
8	1.34	1.65	2.18	2.73	3.49	5.07	7.34	10.2	13.4	15.5	17.5	20.1	22.0	26.1	8
9	1.73	2.09	2.70	3.33	4.17	5.90	8.34	11.4	14.7	16.9	19.0	21.7	23.6	27.9	9
10	2.16	2.56	3.25	3.94	4.87	6.74	9.34	12.5	16.0	18.3	20.5	23.2	25.2	29.6	10
11	2.60	3.05	3.82	4.57	5.58	7.58	10.3	13.7	17.3	19.7	21.9	24.7	26.8	31.3	11
12	3.07	3.57	4.40	5.23	6.30	8.44	11.3	14.8	18.5	21.0	23.3	26.2	28.3	32.9	12
13	3.57	4.11	5.01	5.89	7.04	9.30	12.3	16.0	19.8	22.4	24.7	27.7	29.8	34.5	13
14	4.07	4.66	5.63	6.57	7.79	10.2	13.3	17.1	21.1	23.7	26.1	29.1	31.3	36.1	14
15	4.60	5.23	6.26	7.26	8.55	11.0	14.3	18.2	22.3	25.0	27.5	30.6	32.8	37.2	15
16	5.14	5.81	6.91	7.96	9.31	11.9	15.3	19.4	23.5	26.3	28.8	32.0	34.3	39.3	16
17	5.70	6.41	7.56	8.67	10.1	12.8	16.3	20.5	24.8	27.6	30.2	33.4	35.7	40.8	17
18	6.26	7.01	8.23	9.39	10.9	13.7	17.3	21.6	26.0	28.9	31.5	34.8	37.2	42.3	18
19	6.84	7.63	8.91	10.1	11.7	14.6	18.3	22.7	27.2	30.1	32.9	36.2	38.6	43.8	19
20	7.43	8.26	9.59	10.9	12.4	15.5	19.3	23.8	28.4	31.4	34.2	37.6	40.0	45.3	20
21	8.03	8.90	10.3	11.6	13.2	16.3	20.3	24.9	29.6	32.7	35.5	38.9	41.4	46.8	21
22	8.64	9.54	11.0	12.3	14.0	17.2	21.3	26.0	30.8	33.9	36.8	40.3	42.8	48.3	22
23	9.26	10.2	11.7	13.1	14.8	18.1	22.3	27.1	32.0	35.2	38.1	41.6	44.2	49.7	23
24	9.89	10.9	12.4	13.8	15.7	19.0	23.3	28.2	33.2	36.4	39.4	43.0	45.6	51.2	24
25	10.5	11.5	13.1	14.6	16.5	19.9	24.3	29.3	34.4	37.7	40.6	44.3	46.9	52.6	25
26	11.2	12.2	13.8	15.4	17.3	20.8	25.3	30.4	35.6	38.9	41.9	45.6	48.3	54.1	26
27	11.8	12.9	14.6	16.2	18.1	21.7	26.3	31.5	36.7	40.1	43.2	47.0	49.6	55.5	27
28	12.5	13.6	15.3	16.9	18.9	22.7	27.3	32.6	37.9	41.3	44.5	48.3	51.0	56.9	28
29	13.1	14.3	16.0	17.7	19.8	23.6	28.3	33.7	39.1	42.6	45.7	49.6	52.3	58.3	29
30	13.8	15.0	16.8	18.5	20.6	24.5	29.3	34.8	40.3	43.8	47.0	50.9	53.7	59.7	30

[a] Table A-7 is taken from Fisher and Yates *Statistical Tables for Biological, Agricultural and Medical Research,* published by Oliver & Boyd Ltd., Edinburgh, and by permission of the authors and publishers.

Table A-8 Percentage Points of the Variance Ratio (F-Distribution).[a]

ϕ_2	Per Cent Point	ϕ_1 (corresponding to greater mean square)																			Per Cent Point	ϕ_2
		1	2	3	4	5	6	7	8	9	10	12	15	20	24	30	40	60	120	∞		
1	10	39.9	49.5	53.6	55.8	57.2	58.2	58.9	59.4	59.9	60.2	60.7	61.2	61.7	62.0	62.3	62.5	62.8	63.1	63.3	10	1
	5	161	199	216	225	230	234	237	239	241	242	244	246	248	249	250	251	252	253	254	5	
	1	4052	4999	5403	5625	5764	5859	5928	5982	6022	6056	6106	6157	6209	6235	6261	6287	6313	6339	6366	1	
2	10	8.53	9.00	9.16	9.24	9.29	9.33	9.35	9.37	9.38	9.39	9.41	9.42	9.44	9.45	9.46	9.47	9.47	9.48	9.49	10	2
	5	18.5	19.0	19.2	19.2	19.3	19.3	19.4	19.4	19.4	19.4	19.4	19.4	19.4	19.5	19.5	19.5	19.5	19.5	19.5	5	
	1	98.5	99.0	99.2	99.2	99.3	99.3	99.4	99.4	99.4	99.4	99.4	99.4	99.4	99.5	99.5	99.5	99.5	99.5	99.5	1	
3	10	5.54	5.46	5.39	5.34	5.31	5.28	5.27	5.25	5.24	5.23	5.22	5.20	5.18	5.18	5.17	5.16	5.15	5.14	5.13	10	3
	5	10.1	9.55	9.28	9.12	9.01	8.94	8.89	8.85	8.81	8.79	8.74	8.70	8.66	8.64	8.62	8.59	8.57	8.55	8.53	5	
	1	34.1	30.8	29.5	28.7	28.2	27.9	27.7	27.5	27.3	27.2	27.1	26.9	26.7	26.6	26.5	26.4	26.3	26.2	26.1	1	
4	10	4.54	4.32	4.19	4.11	4.05	4.01	3.98	3.95	3.94	3.92	3.90	3.87	3.84	3.83	3.82	3.80	3.79	3.78	3.76	10	4
	5	7.71	6.94	6.59	6.39	6.26	6.16	6.09	6.04	6.00	5.96	5.91	5.86	5.80	5.77	5.75	5.72	5.69	5.66	5.63	5	
	1	21.2	18.0	16.7	16.0	15.5	15.2	15.0	14.8	14.7	14.5	14.4	14.2	14.0	13.9	13.8	13.7	13.7	13.6	13.5	1	
5	10	4.06	3.78	3.62	3.52	3.45	3.40	3.37	3.34	3.32	3.30	3.27	3.24	3.21	3.19	3.17	3.16	3.14	3.12	3.10	10	5
	5	6.61	5.79	5.41	5.19	5.05	4.95	4.88	4.82	4.77	4.74	4.68	4.62	4.56	4.53	4.50	4.46	4.43	4.40	4.36	5	
	1	16.3	13.3	12.1	11.4	11.0	10.7	10.5	10.3	10.2	10.1	9.89	9.72	9.55	9.47	9.38	9.29	9.20	9.11	9.02	1	
6	10	3.78	3.46	3.29	3.18	3.11	3.05	3.01	2.98	2.96	2.94	2.90	2.87	2.84	2.82	2.80	2.78	2.76	2.74	2.72	10	6
	5	5.99	5.14	4.76	4.53	4.39	4.28	4.21	4.15	4.10	4.06	4.00	3.94	3.87	3.84	3.81	3.77	3.74	3.70	3.67	5	
	1	13.7	10.9	9.78	9.15	8.75	8.47	8.26	8.10	7.98	7.87	7.72	7.56	7.40	7.31	7.23	7.14	7.06	6.97	6.88	1	
7	10	3.59	3.26	3.07	2.96	2.88	2.83	2.78	2.75	2.72	2.70	2.67	2.63	2.59	2.58	2.56	2.54	2.51	2.49	2.47	10	7
	5	5.59	4.74	4.35	4.12	3.97	3.87	3.79	3.73	3.68	3.64	3.57	3.51	3.44	3.41	3.38	3.34	3.30	3.27	3.23	5	
	1	12.2	9.55	8.45	7.85	7.46	7.19	6.99	6.84	6.72	6.62	6.47	6.31	6.16	6.07	5.99	5.91	5.82	5.74	5.65	1	
8	10	3.46	3.11	2.92	2.81	2.73	2.67	2.62	2.59	2.56	2.54	2.50	2.46	2.42	2.40	2.38	2.36	2.34	2.32	2.29	10	8
	5	5.32	4.46	4.07	3.84	3.69	3.58	3.50	3.44	3.39	3.35	3.28	3.22	3.15	3.12	3.08	3.04	3.01	2.97	2.93	15	
	1	11.3	8.65	7.59	7.01	6.63	6.37	6.18	6.03	5.91	5.81	5.67	5.52	5.36	5.28	5.20	5.12	5.03	4.95	4.86	1	
9	10	3.36	3.01	2.81	2.69	2.61	2.55	2.51	2.47	2.44	2.42	2.38	2.34	2.30	2.28	2.25	2.23	2.21	2.18	2.16	10	9
	5	5.12	4.26	3.86	3.63	3.48	3.37	3.29	3.23	3.18	3.14	3.07	3.01	2.94	2.90	2.86	2.83	2.79	2.75	2.71	5	
	1	10.6	8.02	6.99	6.42	6.06	5.80	5.61	5.47	5.35	5.26	5.11	4.96	4.81	4.73	4.65	4.57	4.48	4.40	4.31	1	

Table A-8 (continued)

Per Cent Point	ϕ_2	ϕ_1 (corresponding to greater mean square) 1	2	3	4	5	6	7	8	9	10	12	15	20	24	30	40	60	120	∞	ϕ_2	Per Cent Point
10	10	3.28	2.92	2.73	2.61	2.52	2.46	2.41	2.38	2.35	2.32	2.28	2.24	2.20	2.18	2.16	2.13	2.11	2.08	2.06	10	10
5		4.96	4.10	3.71	3.48	3.33	3.22	3.14	3.07	3.02	2.98	2.91	2.84	2.77	2.74	2.70	2.66	2.62	2.58	2.54		5
1		10.0	7.56	6.55	5.99	5.64	5.39	5.20	5.06	4.94	4.85	4.71	4.56	4.41	4.33	4.25	4.17	4.08	4.00	3.91		1
10	11	3.23	2.86	2.66	2.54	2.45	2.39	2.34	2.30	2.27	2.25	2.21	2.17	2.12	2.10	2.08	2.05	2.03	2.00	1.97	11	10
5		4.84	3.98	3.59	3.36	3.20	3.09	3.01	2.95	2.90	2.85	2.79	2.72	2.65	2.61	2.57	2.53	2.49	2.45	2.40		5
1		9.65	7.21	6.22	5.67	5.32	5.07	4.89	4.74	4.63	4.54	4.40	4.25	4.10	4.02	3.94	3.86	3.78	3.69	3.60		1
10	12	3.18	2.81	2.61	2.48	2.39	2.33	2.28	2.24	2.21	2.19	2.15	2.10	2.06	2.04	2.01	1.99	1.96	1.93	1.90	12	10
5		4.75	3.89	3.49	3.26	3.11	3.00	2.91	2.85	2.80	2.75	2.69	2.62	2.54	2.51	2.47	2.43	2.38	2.34	2.30		5
1		9.33	6.93	5.95	5.41	5.06	4.82	4.64	4.50	4.39	4.30	4.16	4.01	3.86	3.78	3.70	3.62	3.54	3.45	3.36		1
10	13	3.14	2.76	2.56	2.43	2.35	2.28	2.23	2.20	2.16	2.14	2.10	2.05	2.01	1.98	1.96	1.93	1.90	1.88	1.85	13	10
5		4.67	3.81	3.41	3.18	3.03	2.92	2.83	2.77	2.71	2.67	2.60	2.53	2.46	2.42	2.38	2.34	2.30	2.25	2.21		5
1		9.07	6.70	5.74	5.21	4.86	4.62	4.44	4.30	4.19	4.10	3.96	3.82	3.66	3.59	3.51	3.43	3.34	3.25	3.17		1
10	14	3.10	2.73	2.52	2.39	2.31	2.24	2.19	2.15	2.12	2.10	2.05	2.01	1.96	1.94	1.91	1.89	1.86	1.83	1.80	14	10
5		4.60	3.74	3.34	3.11	2.96	2.85	2.76	2.70	2.65	2.60	2.53	2.46	2.39	2.35	2.31	2.27	2.22	2.18	2.13		5
1		8.86	6.51	5.56	5.04	4.69	4.46	4.28	4.14	4.03	3.94	3.80	3.66	3.51	3.43	3.35	3.27	3.18	3.09	3.00		1
10	15	3.07	2.70	2.49	2.36	2.27	2.21	2.16	2.12	2.09	2.06	2.02	1.97	1.92	1.90	1.87	1.85	1.82	1.79	1.76	15	10
5		4.54	3.68	3.29	3.06	2.90	2.79	2.71	2.64	2.59	2.54	2.48	2.40	2.33	2.29	2.25	2.20	2.16	2.11	2.07		5
1		8.68	6.36	5.42	4.89	4.56	4.32	4.14	4.00	3.89	3.80	3.67	3.52	3.37	3.29	3.21	3.13	3.05	2.96	2.87		1
10	16	3.05	2.67	2.46	2.33	2.24	2.18	2.13	2.09	2.06	2.03	1.99	1.94	1.89	1.87	1.84	1.81	1.78	1.75	1.72	16	10
5		4.49	3.63	3.24	3.01	2.85	2.74	2.66	2.59	2.54	2.49	2.42	2.35	2.28	2.24	2.19	2.15	2.11	2.06	2.01		5
1		8.53	6.23	5.29	4.77	4.44	4.20	4.03	3.89	3.78	3.69	3.55	3.41	3.26	3.18	3.10	3.02	2.93	2.84	2.75		1
10	17	3.03	2.64	2.44	2.31	2.22	2.15	2.10	2.06	2.03	2.00	1.96	1.91	1.86	1.84	1.81	1.78	1.75	1.72	1.69	17	10
5		4.45	3.59	3.20	2.96	2.81	2.70	2.61	2.55	2.49	2.45	2.38	2.31	2.23	2.19	2.15	2.10	2.06	2.01	1.96		5
1		8.40	6.11	5.18	4.67	4.34	4.10	3.93	3.79	3.68	3.59	3.46	3.31	3.16	3.08	3.00	2.92	2.83	2.75	2.65		1
10	18	3.01	2.62	2.42	2.29	2.20	2.13	2.08	2.04	2.00	1.98	1.93	1.89	1.84	1.81	1.78	1.75	1.72	1.69	1.66	18	10
5		4.41	3.55	3.16	2.93	2.77	2.66	2.58	2.51	2.46	2.41	2.34	2.27	2.19	2.15	2.11	2.06	2.02	1.97	1.92		5
1		8.29	6.01	5.09	4.58	4.25	4.01	3.84	3.71	3.60	3.51	3.37	3.23	3.08	3.00	2.92	2.84	2.75	2.66	2.57		1

Table A-8 (continued)

ϕ_2	Per Cent Point	ϕ_1 (corresponding to greater mean square)																		
		1	2	3	4	5	6	7	8	9	10	12	15	20	24	30	40	60	120	∞
19	10	2.99	2.61	2.40	2.27	2.18	2.11	2.06	2.02	1.98	1.96	1.91	1.86	1.81	1.79	1.76	1.73	1.70	1.67	1.63
	5	4.38	3.52	3.13	2.90	2.74	2.63	2.54	2.48	2.42	2.38	2.31	2.23	2.16	2.11	2.07	2.03	1.98	1.93	1.88
	1	8.18	5.93	5.01	4.50	4.17	3.94	3.77	3.63	3.52	3.43	3.30	3.15	3.00	2.92	2.84	2.76	2.67	2.58	2.49
20	10	2.97	2.59	2.38	2.25	2.16	2.09	2.04	2.00	1.96	1.94	1.89	1.84	1.79	1.77	1.74	1.71	1.68	1.64	1.61
	5	4.35	3.49	3.10	2.87	2.71	2.60	2.51	2.45	2.39	2.35	2.28	2.20	2.12	2.08	2.04	1.99	1.95	1.90	1.84
	1	8.10	5.85	4.94	4.43	4.10	3.87	3.70	3.56	3.46	3.37	3.23	3.09	2.94	2.86	2.78	2.69	2.61	2.52	2.42
21	10	2.96	2.57	2.36	2.23	2.14	2.08	2.02	1.98	1.95	1.92	1.87	1.83	1.78	1.75	1.72	1.69	1.66	1.62	1.59
	5	4.32	3.47	3.07	2.84	2.68	2.57	2.49	2.42	2.37	2.32	2.25	2.18	2.10	2.05	2.01	1.96	1.92	1.87	1.81
	1	8.02	5.78	4.87	4.37	4.04	3.81	3.64	3.51	3.40	3.31	3.17	3.03	2.88	2.80	2.72	2.64	2.55	2.46	2.36
22	10	2.95	2.56	2.35	2.22	2.13	2.06	2.01	1.97	1.93	1.90	1.86	1.81	1.76	1.73	1.70	1.67	1.64	1.60	1.57
	5	4.30	3.44	3.05	2.82	2.66	2.55	2.46	2.40	2.34	2.30	2.23	2.15	2.07	2.03	1.98	1.94	1.89	1.84	1.78
	1	7.95	5.72	4.82	4.31	3.99	3.76	3.59	3.45	3.35	3.26	3.12	2.98	2.83	2.75	2.67	2.58	2.50	2.40	2.31
23	10	2.94	2.55	2.34	2.21	2.11	2.05	1.99	1.95	1.92	1.89	1.85	1.80	1.74	1.72	1.69	1.66	1.62	1.59	1.55
	5	4.28	3.42	3.03	2.80	2.64	2.53	2.44	2.37	2.32	2.27	2.20	2.13	2.05	2.00	1.96	1.91	1.86	1.81	1.76
	1	7.88	5.66	4.76	4.26	3.94	3.71	3.54	3.41	3.30	3.21	3.07	2.93	2.78	2.70	2.62	2.54	2.45	2.35	2.26
24	10	2.93	2.54	2.33	2.19	2.10	2.04	1.98	1.94	1.91	1.88	1.83	1.78	1.73	1.70	1.67	1.64	1.61	1.57	1.53
	5	4.26	3.40	3.01	2.78	2.62	2.51	2.42	2.36	2.30	2.25	2.18	2.11	2.03	1.98	1.94	1.89	1.84	1.79	1.73
	1	7.82	5.61	4.72	4.22	3.90	3.67	3.50	3.36	3.26	3.17	3.03	2.89	2.74	2.66	2.58	2.49	2.40	2.31	2.21
25	10	2.92	2.53	2.32	2.18	2.09	2.02	1.97	1.93	1.89	1.87	1.82	1.77	1.72	1.69	1.66	1.63	1.59	1.56	1.52
	5	4.24	3.39	2.99	2.76	2.60	2.49	2.40	2.34	2.28	2.24	2.16	2.09	2.01	1.96	1.92	1.87	1.82	1.77	1.71
	1	7.77	5.57	4.68	4.18	3.86	3.63	3.46	3.32	3.22	3.13	2.99	2.85	2.70	2.62	2.54	2.45	2.36	2.27	2.17
26	10	2.91	2.52	2.31	2.17	2.08	2.01	1.96	1.92	1.88	1.86	1.81	1.76	1.71	1.68	1.65	1.61	1.58	1.54	1.50
	5	4.23	3.37	2.98	2.74	2.59	2.47	2.39	2.32	2.27	2.22	2.15	2.07	1.99	1.95	1.90	1.85	1.80	1.75	1.69
	1	7.72	5.53	4.64	4.14	3.82	3.59	3.42	3.29	3.18	3.09	2.96	2.82	2.66	2.58	2.50	2.42	2.33	2.23	2.13
27	10	2.90	2.51	2.30	2.17	2.07	2.00	1.95	1.91	1.87	1.85	1.80	1.75	1.70	1.67	1.64	1.60	1.57	1.53	1.49
	5	4.21	3.35	2.96	2.73	2.57	2.46	2.37	2.31	2.25	2.20	2.13	2.06	1.97	1.93	1.88	1.84	1.79	1.73	1.67
	1	7.68	5.49	4.60	4.11	3.78	3.56	3.39	3.26	3.15	3.06	2.93	2.78	2.63	2.55	2.47	2.38	2.29	2.20	2.10

Table A-8 (concluded)

Per Cent Point	ϕ_2	ϕ_1 (corresponding to greater mean square)																			ϕ_2	Per Cent Point
		1	2	3	4	5	6	7	8	9	10	12	15	20	24	30	40	60	120	∞		
10	28	2.89	2.50	2.29	2.16	2.06	2.00	1.94	1.90	1.87	1.84	1.79	1.74	1.69	1.66	1.63	1.59	1.56	1.52	1.48	28	10
5		4.20	3.34	2.95	2.71	2.56	2.45	2.36	2.29	2.24	2.19	2.12	2.04	1.96	1.91	1.87	1.82	1.77	1.71	1.65		5
1		7.64	5.45	4.57	4.07	3.75	3.53	3.36	3.23	3.12	3.03	2.90	2.75	2.60	2.52	2.44	2.35	2.26	2.17	2.06		1
10	29	2.89	2.50	2.28	2.15	2.06	1.99	1.93	1.89	1.86	1.83	1.78	1.73	1.68	1.65	1.62	1.58	1.55	1.51	1.47	29	10
5		4.18	3.33	2.93	2.70	2.55	2.43	2.35	2.28	2.22	2.18	2.10	2.03	1.94	1.90	1.85	1.81	1.75	1.70	1.64		5
1		7.60	5.42	4.54	4.04	3.73	3.50	3.33	3.20	3.09	3.00	2.87	2.73	2.57	2.49	2.41	2.33	2.23	2.14	2.03		1
10	30	2.88	2.49	2.28	2.14	2.05	1.98	1.93	1.88	1.85	1.82	1.77	1.72	1.67	1.64	1.61	1.57	1.54	1.50	1.46	30	10
5		4.17	3.32	2.92	2.69	2.53	2.42	2.33	2.27	2.21	2.16	2.09	2.01	1.93	1.89	1.84	1.79	1.74	1.68	1.62		5
1		7.56	5.39	4.51	4.02	3.70	3.47	3.30	3.17	3.07	2.98	2.84	2.70	2.55	2.47	2.39	2.30	2.21	2.11	2.01		1
10	40	2.84	2.44	2.23	2.09	2.00	1.93	1.87	1.83	1.79	1.76	1.71	1.66	1.61	1.57	1.54	1.51	1.47	1.42	1.38	40	10
5		4.08	3.23	2.84	2.61	2.45	2.34	2.25	2.18	2.12	2.08	2.00	1.92	1.84	1.79	1.74	1.69	1.64	1.58	1.51		5
1		7.31	5.18	4.31	3.83	3.51	3.29	3.12	2.99	2.89	2.80	2.66	2.52	2.37	2.29	2.20	2.11	2.02	1.92	1.80		1
10	60	2.79	2.39	2.18	2.04	1.95	1.87	1.82	1.77	1.74	1.71	1.66	1.60	1.54	1.51	1.48	1.44	1.40	1.35	1.29	60	10
5		4.00	3.15	2.76	2.53	2.37	2.25	2.17	2.10	2.04	1.99	1.92	1.84	1.75	1.70	1.65	1.59	1.53	1.47	1.39		5
1		7.08	4.98	4.13	3.65	3.34	3.12	2.95	2.82	2.72	2.63	2.50	2.35	2.20	2.12	2.03	1.94	1.84	1.73	1.60		1
10	120	2.75	2.35	2.13	1.99	1.90	1.82	1.77	1.72	1.68	1.65	1.60	1.54	1.48	1.45	1.41	1.37	1.32	1.26	1.19	120	10
5		3.92	3.07	2.68	2.45	2.29	2.18	2.09	2.02	1.96	1.91	1.83	1.75	1.66	1.61	1.55	1.50	1.43	1.35	1.25		5
1		6.85	4.79	3.95	3.48	3.17	2.96	2.79	2.66	2.56	2.47	2.34	2.19	2.03	1.95	1.86	1.76	1.66	1.53	1.38		1
10	∞	2.71	2.30	2.08	1.94	1.85	1.77	1.72	1.67	1.63	1.60	1.55	1.49	1.42	1.38	1.34	1.30	1.24	1.17	1.00	∞	10
5		3.84	3.00	2.60	2.37	2.21	2.10	2.01	1.94	1.88	1.83	1.75	1.67	1.57	1.52	1.46	1.39	1.32	1.22	1.00		5
1		6.63	4.61	3.78	3.32	3.02	2.80	2.64	2.51	2.41	2.32	2.18	2.04	1.88	1.79	1.70	1.59	1.47	1.32	1.00		1

a Table A-8 is taken from Fisher and Yates *Statistical Tables for Biological, Agricultural and Medical Research,* published by Oliver & Boyd Ltd., Edinburgh, and by permission of the authors and publishers.

INDEX

INDEX

Abscissa, 67
Accuracy, definition, 145
Addition, rule, 97
Arithmetic mean, 2
Average, 1

Bias, 7, 20
Binominal distribution, 106
   mean of, 106
   standard deviation and variance of, 106
Binomial expansion, 103, 107
Bravais-Pearson coefficient of correlation, 86

Cardinal numbers, 87
Charts, control, 146
Chi-square, 116
   distribution, 117
Class boundaries, 23
Classification of data, 22
Coefficient of correlation, 86, 94
   rank correlation, 88, 93
Compound probability, 98
Confidence limits, 54
   description of, 49
   estimation from paired replicates, 151
   of mean, 49,
   of proportions, 103, 108
   of standard deviation and variance, 118, 124
Continuous variables, 3, 5
Contingency tables, 120
   degrees of freedom in, 121
Control charts, 146
Coordinates, 67
Correlation coefficient, 86, 88, 93, 94
Cumulative frequency, 26

Degrees of freedom, 52, 53, 54
   in analysis of variance, 136, 138, 139, 140, 143
   in Chi-square test, 119
   in F test, 123
De Moivre, 17
Dependent variables, 68
Discontinuous variables, 35

Error, law, 29
   standard, of estimate, 82
   standard, of mean, 48, 54

F-test, 123, 124
   in analysis of variance, 137, 139
Fiducial limit, 49
Frequency distribution, 15
   polygon, 104, 110

Galton, 76
Gauss, 17, 29
Gaussian distribution (see Normal Distribution)
Geometric mean, 33, 43
Gosset, W.S., 52

Harmonic mean, 32, 43
Histogram, 16, 27, 12, 9 (see also Frequency polygon)
Hyperbola, 74
Hypothesis, null, 55, 56, 63, 88

Inference, 5, 9

Laplace, 17
Least squares, 80
Linear equations, 69, 76

Mean(s), arithmetic, 2, 27, 37, 43
    comparison of, 64, 131
    confidence limits, 49, 50
    distributions of, 47
    geometric, 2, 33, 43
    harmonic, 2, 43
    standard error of, 48, 54
    universe, 47, 48, 54, 137
Median, 27, 28, 34, 163
    confidence limits, 164
Mode, 27, 34
Multiplication principle or rule, 12, 95,
    97, 98
Mutually exclusive events, 98

Normal distribution, 2, 17, 29
    equation of, 34
Null hypothesis, 55, 56, 63, 88

Ordinal numbers, 88
Ordinate, 67

Paired observations, 58
Parabola, 74
Parameters, 34, 37
Pascal's triangle, 14
Point binomal, 103, 111
Point Poisson, 115
Poisson distribution, 112
    mean and variance of, 114
Polygon, frequency, 104, 110
Population(s), 9, 46
Precision, definition, 145
Prediction, 9
Probability, 4, 5
    compound, 98
Proportion(s), 95, 103
    comparisons of, 121, 125
    confidence limits, 109

Random, 6, 7, 18
Random numbers, 20
Randomization, 20
Rank correlation coefficient, 88, 93
Regression, 76
Residual, 143

Sample size, 46, 160
Sampling, random, 9, 10, 20
Sign test, 163, 165
Significance, 55
Slope, 71
Standard deviation, 34, 35, 37, 38, 44
    computation of, 38
    comparisons of (see F-test)
    conficence limits, 118
Standard error of estimate, 82
Standard error of mean, 48, 54
Statistics, definition, 4, 5
"Student", 52
Snedecor's F Distribution, 123

t, distribution, 52
    test, 52, 54, 58, 131

Universe of discourse, 46
Universe mean and standard deviation, 47

Variable(s), continuous and discontinuous,
    3, 5
    dependent, 68
    independent, 68
Variance, 38, 43
    pooled, 62

Yate's correction, 122